TERNARY DIAMOND-LIKE SEMICONDUCTORS

TROINYE ALMAZOPODOBNYE POLUPROVODNIKI

ТРОЙНЫЕ АЛМАЗОПОДОБНЫЕ ПОЛУПРОВОДНИКИ

TERNARY DIAMOND-LIKE SEMICONDUCTORS

Lev I. Berger

All-Union Institute for Chemical Reagents and Very Pure Substances
Moscow, USSR

and

Vitalii D. Prochukhan

A. F. Ioffe Physicotechnical Institute,
Academy of Sciences of the USSR
Leningrad, USSR

Translated from Russian by
Albin Tybulewicz
Editor, *Soviet Physics-Semiconductors*

CONSULTANTS BUREAU · NEW YORK – LONDON · 1969

Lev Isaakovich Berger was born in 1929. He obtained his first degree at the Physico-Mathematical Department of the State Correspondence Pedagogical Institute in Moscow. In 1959, Berger defended a thesis on "Thermal conductivity and thermal expansion of some semiconducting compounds," for which he was awarded the degree of Candidate of Physico-Mathematical Sciences. In 1968, he offered a thesis on "Investigations of ternary diamond-like semiconductors," which won him the degree of Doctor of Technical Sciences. He is presently head of the Semiconducting Materials Section at the All-Union Scientific Research Institute for Chemical Reagents and Very Pure Substances

Vitalii Danilovich Prochukhan was born in 1931. He graduated from the Chemistry Department of the Leningrad State University and is presently a senior scientist at the A.F. Ioffe Physicotechnical Institute of the Academy of Sciences of the USSR. In 1962, Prochukhan defended a thesis on "Some multicomponent semiconducting phases," for which he obtained the degree of Candidate of Chemical Sciences.

The original Russian text, published by Metallurgiya Press in Moscow in 1968, has been corrected by the authors for this edition

БЕРГЕР *Лев Исаакович*   ПРОЧУХАН *Виталий Данилович*
ТРОЙНЫЕ АЛМАЗОПОДОВНЫЕ
ПОЛУПРОВОДНИКИ

ISBN 978-1-4757-0042-8   ISBN 978-1-4757-0040-4 (eBook)
DOI 10.1007/978-1-4757-0040-4

Library of Congress Catalog Card Number 69-17903

© 1969 Consultants Bureau, New York
A Division of Plenum Publishing Corporation
227 West 17th Street, New York, N. Y. 10011

United Kingdom edition published by Consultants Bureau, London
A Division of Plenum Publishing Company, Ltd.
Donington House, 30 Norfolk Street, London, W.C. 2, England

CONTENTS

PREFACE

The science of semiconducting materials is still very young. Less than twenty years ago, the school of A. F. Ioffe demonstrated that the properties of semiconductors are governed primarily by their chemical nature and can be predicted on this basis. These ideas are still being developed and used to establish a new materials science: the chemistry of semiconductors.

The solution of problems in the chemistry of semiconductors should make it possible to find new applications for solids. We are already witnessing the process in which the practical importance of such new materials as diamond-like $A^{III}B^V$ compounds is accelerating the development of the chemistry and physics of semiconductors and some allied sciences.

Diamond-like semiconductors are promising materials for modern electronics. They belong to an extensive class of valence compounds which seem to be an inexhaustible source of new semiconducting materials.

Among these new, particularly promising materials, are ternary diamond-like semiconductors, which are the subject of the present monograph. The appearance of this book, which is the first on this subject not only in the Soviet Union but also outside it, is the proof of the importance attached to promising semiconductors in the USSR.

The authors describe the methods for the preparation of compounds and the growth of single crystals. They analyze in detail the physicochemical and physical properties of ternary compounds and the relationships between these properties, and consider the possible applications of these substances and suggest further investigations.

The introductory chapter, written by the editor, shows the place of ternary diamond-like semiconducting compounds among semiconducting materials, and considers the possibilities of predicting semiconducting properties in such compounds.

The book should be useful to both chemists and physicists working on semiconducting materials and devices.

N. A. Goryunova (Editor)

CHAPTER I

TERNARY DIAMOND-LIKE COMPOUNDS AND THEIR POSITION AMONG OTHER SEMICONDUCTING SUBSTANCES

System of Valence Compounds

Much has been published on diamond-like semiconducting materials.

The interest in these materials has been generated by the important applications of many members of this group of compounds, and the theory of semiconductors has been extended considerably by the use of models based on the chemical bonds in these materials.

Ternary compounds occupy a special place among diamond-like materials. After the establishment of the principal properties of $A^{III}B^V$ and $A^{II}B^{VI}$ semiconductors, as well as the properties of solid solutions based on these compounds, it became clear that the next stage of the search for new semiconductors should be directed toward heterovalent ternary compounds. Ternary diamond-like compounds are very close electron-nuclear analogs of $A^{III}B^V$ and $A^{II}B^{VI}$ compounds in respect of their composition, crystal structure, and nature of chemical bonds. However, in addition to their similarities with these binary compounds, ternary diamond-like materials also exhibit considerable differences, and it is these differences that have attracted investigators to these fairly complex materials.

One of the most important properties of the structure of ternary compounds is the possibility of an additional ordering compared with the structure of binary diamond-like compounds. In the sphalerite and wurtzite structures of $A^{III}B^V$ and $A^{II}B^{VI}$ compounds, cations are ordered relative to anions (a cation is surrounded only by anions and vice versa), while in ternary compounds there are two different cations (or anions), and therefore we can have the ordering of cations relative to one another.

Many ternary compounds crystallize in the tetragonal structure of chalcopyrite. This means that the lattice symmetry is lower than that in a binary compound and that there should be a corresponding change in the energy spectrum. We may assume that such a change can affect considerably the mechanism of charge transport in these semiconductors. This point is undoubtedly of interest because we are dealing with substances which are closer to $A^{III}B^V$ and $A^{II}B^{VI}$ compounds than any others.

The physicochemical properties of ternary compounds have also attracted attention. In spite of the considerable complexity of the processes of synthesis of ternary compounds, compared with binary materials, the earliest investigations have already established that ternary compounds have many advantages over binary ones, particularly in respect of their much lower melting points. When the applications of such compounds are considered, the low melting point not only makes it easier to prepare them, but provides an opportunity to reduce the contamination during synthesis.

These considerations, together with the search for new semiconductors whose properties could be interpreted using the model of charge transport developed for $A^{III}B^{V}$ and $A^{II}B^{VI}$ compounds, make investigations of the physical and physicochemical properties of ternary diamond-like compounds extremely interesting.

Three questions arise: which types of ternary compounds should we consider; is the number of these compounds limited; and what is their position in the class of valence semiconductors, which include diamond-like phases?

We shall consider first the classification of semiconducting materials.

The idea of considering jointly both semiconductors and dielectrics is not new to solid-state physics. It can now be regarded as fully accepted. The band theory of solids has shown that the energy spectra of all inorganic compounds exhibiting nonmetallic conduction, including ionic compounds, are basically similar and that the formation of a forbidden band is related to the completion of electron shells in the components of a compound.

These developments in solid-state physics occurred at approximately the same time as the evolution of the concept of ionic-covalent binding in the quantum chemistry. The term "valence" acquired a new meaning. Thus, in a recent book [17], Spice gives the following definitions of valence: a) the number of electrons which an atom loses or acquires when it forms a simple ion; b) the number of double (two-electron) bonds formed by a neutral atom. From Spice's point of view, semiconductors are valence compounds, all or some of whose atoms acquire a complete-shell electron structure by the formation of two-electron bonds or by the transfer of valence electrons from a cation to an anion.

However, not all semiconductors can be described on the basis of the ionic—covalent interaction of atoms and, moreover, this interaction itself has no satisfactory quantum-chemical model. Syrkin [300] draws attention to the need to consider, in addition to the main factors, such as localized molecular orbits and ionic bonds, the possible presence of empty orbits, unshared electron pairs, donor—acceptor and dative bonds, various types of hybridization, etc.

Goryunova [326] has pointed out that many semiconducting materials have a very complex pattern of chemical bonds, particularly in the presence of atoms with incomplete valence, or atoms of transition elements. There are, however, simpler groups of chemical compounds which are formed in accordance with simple valence laws. They are the compounds, discovered in the nineteen thirties by Goldschmidt, whose electron structure is similar to that of diamond, as well as the compounds with the sodium chloride structure or with the electron configurations of inert gases, whose formation characteristics were discussed many years ago by Kossel.

These groups of tetrahedral and octahedral phases, with an average of four electrons per atom, form the main part of the class valence chemical compounds, which have a simpler ionic—covalent interaction than other substances: complete electron shells are formed by all atoms in the compound using all the valence electrons. Therefore, we may expect that the semiconducting properties, i.e., the energy gap between the conduction and valence bands, should be a characteristic property of all valence chemical compounds. The same conclusion is reached by means of various empirical formulas, used in the prediction of semiconducting properties. These formulas are based on the assumption of the formation of complete electron shells, which is the most characteristic property of the class of valence compounds.

When we compare typical formulas of all possible binary valence compounds with the component elements, we find that the four-electron tetrahedral and octahedral phases are located between nonequiatomic compounds in which the number of electrons per atom is not equal to four (Table 1). It is evident from this table that these phases form the central core of the class of valence compounds.

TABLE 1. Binary Valence Compounds

General formula	Typical element or compound	No. of electrons per atom	General formula	Typical element or compound	No. of electrons per atom
A^{I}	Cu, K	1	$A^{II} B^{VI}$	ZnSe; CaSe	4.00
$A^{I}_6 B^{II}$	Cu_6 Zn; K_6Ca	1.14	$A^{III} B^{V}$	GaAs; ScAs	4.00
$A^{I}_5 B^{III}$	Cu_5Ga; K_5Ga	1.33	$A^{IV}_3 B^{V}_4$	Ge_3As_4; Ti_3As_4	4.57
$A^{I}_4 B^{IV}$	Cu_4Ge; K_4Ge	1.60	$A^{III}_2 B^{VI}_3$	Ga_2Se_3; Sc_2Se_3	4.80
A^{II}	Zn; Ca	2.00	A^{V}	As	5.00
$A^{I}_3 B^{V}$	Cu_3As; K_3As	2.00	$A^{II} B^{VI}_2$	$ZnBr_2$; $CaBr_2$	5.33
$A^{II}_5 B^{III}_2$	Zn_5 Ga_2; Ca_5Ga_2	2.29	$A^{IV} B^{VI}_2$	$GeSe_2$; $TiSe_2$	5.33
$A^{I}_2 B^{VI}$	Cu_2Se; K_2Se	2.67	$A^{V}_2 B^{VI}_5$	As_2Se_5	5.71
$A^{II}_2 B^{IV}$	Zn_2Ge; Ca_2Ge	2.67	A^{VI}	Se	6.00
A^{3}	Ga	3.00	$A^{III} B^{VII}_3$	$GaBr_3$; $ScBr_3$	6.00
$A^{II}_3 B^{V}_2$	Zn_3As_2; Ca_3As_2	3.20	$A^{IV} B^{VII}_4$	$GeBr_4$; $TiBr_4$	6.40
$A^{III}_4 B^{IV}_3$	Ga_4Ge_3; Sc_4Ge_3	3.43	$A^{V} B^{VII}_5$	$AsBr_5$	6.67
A^{IV}	Ge	4.00	$A^{VI} B^{VII}_6$	$SeBr_6$	6.86
$A^{I} B^{VII}$	CuBr; KBr	4.00	A^{VII}	Br	7.00

Four-electron compounds include such well-known semiconductors as germanium, silicon, and $A^{III}B^{V}$ and $A^{II}B^{VI}$ compounds. However, experimental evidence shows that semiconducting properties are exhibited also by substances in which the number of electrons per atom differs from four. Thus, some elements in the third (boron), fifth (phosphorus, arsenic, and antimony), sixth (sulfur, selenium, and tellurium), as well as the seventh (iodine) groups of the periodic table also exhibit semiconducting properties. Semiconducting materials also include binary nonequiatomic compounds such as $A^{II}_2 B^{IV}$ or $A^{II}_3 B^{V}_2$ compounds (for example, Mg_2Si and Zn_3As_2).

The question now arises of the limits of the variation of the number of electrons per atom within which we can expect semiconducting properties. These properties have not been observed in elements with two valence electrons. However, binary compounds with two elec-

TABLE 2. Ternary Analogs of Binary Valence Compounds

Formula of binary prototype	Electrons per atom	I–II–III	I–II–IV	I–II–V	I–II–VI	I–II–VII	I–III–IV	I–III–V	I–III–VI	I–III–VII	I–IV–V	I–IV–VI	I–IV–VII
$A_3^I B^V$	2.0	$A^I B_4^{II} C^{III}$	$A_2^I B^{II} C^{IV}$	$A_3^I B^V$	$A_{12}^I B_2^{II} C_3^{VI}$	$A_{15}^I B_2^{II} C_3^{VII}$	$A_6^I B^{III} C_2^{IV}$	$A_3^I B^V$	$A_9^I B^{III} C_2^{VI}$	$A_6^I B^{III} C^{VII}$	$A_3^I B^V$	$A_6^I B^{IV} C^{VI}$	$A_9^I B_2^{IV} C^{VII}$
$A_5^{II} B_2^{III}$	2.29	$A_5^{II} B_2^{III}$	$A_2^I B_3^{II} C_2^{IV}$	$A_4^I B^{II} C_2^V$	$A_{20}^I B^{II} C_7^{VI}$	$A_{25}^I B_3^{II} C_7^{VII}$	$A_7^I B_3^{III} C_4^{IV}$	$A_9^I B^{III} C_4^V$	$A_{15}^I B^{III} C_5^{VI}$	$A_{20}^I B_3^{III} C_5^{VII}$	$A_{14}^I B^{IV} C_6^V$	$A_{10}^I B^{IV} C_3^{VI}$	$A_5^I B^{IV} C^{VII}$
$A_2^I B^{VI}$; $A_2^{II} B^{IV}$	2.67	—	$A_2^{II} B^{IV}$	$A^I B^{II} C^V$	$A_2^I B^{VI}$	$A_{10}^I B^{II} C_4^{VII}$	$A^I B^{III} C^{IV}$	$A_3^I B^{III} C_2^V$	$A_2^I B^{VI}$	$A_8^I B^{III} C_3^{VII}$	$A_5^I B^{IV} C_3^V$	$A_2^I B^{VI}$	$A_6^I B^{IV} C_2^{VII}$
$A_3^{II} B_2^V$	3.20	—	—	$A_3^{II} B_2^V$	$A_2^I B^{II} C_2^{VI}$	$A_{15}^I B^{II} C_9^{VII}$	$A^I B_5^{III} C_4^{IV}$	$A_3^I B_3^{III} C_4^V$	$A_5^I B^{III} C_4^{VI}$	$A_{12}^I B^{III} C_7^{VII}$	$A_2^I B^{IV} C_2^V$	$A_8^I B^{IV} C_6^{VI}$	$A_9^I B^{IV} C_5^{VII}$
$A_3^{III} B_2^V$	3.42	—	—	—	$A_2^I B_2^{II} C_3^{VI}$	$A_{20}^I B^{II} C_{14}^{VII}$	$A_4^{III} B_3^{IV}$	$A_3^I B_5^{III} C_6^V$	$A_3^I B^{III} C_3^{VI}$	$A_{16}^I B^{III} C_{11}^{VII}$	$A_7^I B^{IV} C_5^V$	$A_{10}^I B_2^{IV} C_9^{VI}$	$A_{12}^I B^{IV} C_8^{VII}$
$A^{III} B^V$; $A^{II} B^{VI}$; $A^I B^{VII}$	4.00	—	—	—	$A^{II} B^{VI}$	$A^I B^{VII}$	—	$A^{III} B^V$	$A^I B^{III} C_2^{VI}$	$A^I B^{VII}$	$A^I B_2^{IV} C_3^V$	$A_2^I B^{IV} C_3^{VI}$	$A^I B^{VII}$
$A_3^{IV} B_4^V$	4.57	—	—	—	—	$A_2^I B^{II} C_4^{VII}$	—	—	$A^I B_5^{III} C_8^{VI}$	$A_5^I B^{III} C_8^{VII}$	$A_3^{IV} B_4^V$	$A_4^I B_5^{IV} C_{12}^{VI}$	$A_8^I B^{IV} C_{12}^{VII}$
$A_2^{III} B_3^{VI}$	4.80	—	—	—	—	$A^I B^{II} C_3^{VII}$	—	—	$A_2^{III} B_3^{VI}$	$A_3^I B^{III} C_6^{VII}$	—	$A_2^I B_4^{IV} C_9^{VI}$	$A_5^I B^{IV} C_9^{VII}$
$A^{II} B_2^{VII}$; $A^{IV} B_2^{VI}$	5.33	—	—	—	—	$A^{II} B_2^{VII}$	—	—	—	$A^I B^{III} C_4^{VII}$	—	$A^{IV} B_2^{VI}$	$A_2^I B^{IV} C_6^{VII}$
$A_2^V B_3^{VI}$	5.71	—	—	—	—	—	—	—	—	$A^I B_3^{III} C_{10}^{VII}$	—	—	$A^I B^{IV} C_5^{VII}$
$A^{III} B_3^{VII}$	6.00	—	—	—	—	—	—	—	—	$A^{III} B_3^{VII}$	—	—	$A^I B_2^{IV} C_9^{VII}$

TABLE 2 (Continued)

Formula of binary prototype	Electrons per atom	I–V–VI	I–V–VII	I–VI–VII	II–III–IV	II–III–V	II–III–VI	II–III–VII	II–IV–V	II–IV–VI	II–IV–VII	II–V–VI	II–V–VII
$A_3^I B^V$	2.0	$A_3^I B^V$	$A_3^I B^V$	—	—	—	—	—	—	—	—	—	—
$A_5^{II} B_2^{III}$	2.29	$A_5^I B^V C_8^{VI}$	$A_{10}^I B_3^V C^{VII}$	—	$A_5^{II} B_2^{III}$	$A_5^{II} B_2^{III}$	$A_5^{II} B_2^{III}$	$A_5^{II} B_2^{III}$	—	—	—	—	—
$A_2^I B^{VI}$; $A_2^{II} B^{IV}$	2.67	$A_2^I B^{VI}$	$A_4^I B^V C^{VII}$	$A_2^I B^{VI}$	$A_2^{II} B^{IV}$	$A_4^{II} B^{III} C^V$	$A_6^{II} B_2^{III} C^{VI}$	—	$A_2^{II} B^{IV}$	$A_2^{II} B^{IV}$	$A_2^{II} B^{IV}$	—	—
$A_3^{II} B_2^V$	3.20	$A_{11}^I B^V C_8^{VI}$	$A_6^I B^V C_3^{VII}$	$A_3^I B^{VI} C^{VII}$	$A^{II} B_2^{III} C_2^{IV}$	$A_3^{II} B_2^V$	$A_9^{II} B_2^{III} C_4^{VI}$	$A_3^{II} B^{III} C^{VII}$	$A_3^{II} B_2^V$	$A_3^{II} B^{IV} C^{VI}$	$A_9^{II} B_4^{IV} C_2^{VII}$	$A_3^{II} B_2^V$	$A_3^{II} B_2^V$
$A_4^{III} B_3^{IV}$	3.42	$A_7^I B^V C_6^{VI}$	$A_8^I B^V C_5^{VII}$	$A_4^I B^{VI} C_2^{VII}$	$A_4^{III} B_3^{IV}$	$A_3^{II} B^{III} C_3^V$	$A_{12}^{II} B_2^{III} C_7^{VI}$	$A_{16}^{II} B_5^{III} C_7^{VII}$	$A_7^{II} B^{IV} C_6^V$	$A_4^{II} B^{IV} C_2^{VI}$	$A_{12}^{II} B_5^{IV} C_4^{VII}$	$A_4^{II} B_2^V C^{VI}$	$A_8^{II} B_5^V C^{VII}$
$A^{III} B^V$; $A^{II} B^{VI}$; $A^I B^{VII}$	4.00	$A_3^I B^V C_4^{VI}$	$A^I B^{VII}$	$A^I B^{VII}$	—	$A^{III} B^V$	$A^{II} B^{VI}$	$A_4^{II} B^{III} C_3^{VII}$	$A^{II} B^{IV} C_2^V$	$A^{II} B^{VI}$	$A_3^{II} B^{IV} C_2^{VII}$	$A^{II} B^{VI}$	$A_2^{II} B^V C^{VII}$
$A_3^{IV} B_4^V$	4.57	$A_7^I B_5^V C_{16}^{VI}$	$A_{11}^I B^V C_{16}^{VII}$	$A_{14}^I B^{VI} C_{20}^{VII}$	—	—	$A_2^{II} B^{III} C_4^{VI}$	$A_6^{II} B^{III} C_7^{VII}$	$A_3^{IV} B_4^V$	$A_2^{II} B^{IV} C_4^{VI}$	$A_9^{II} B_2^{IV} C_{10}^{VII}$	$A_7^{II} B_2^V C_{12}^{VI}$	$A_3^{II} B^V C_3^{VII}$
$A_2^{III} B_3^{VI}$	4.80	$A^I B^V C_3^{VI}$	$A_7^I B^V C_{12}^{VII}$	$A_9^I B^{VI} C_{15}^{VII}$	—	—	$A_2^{III} B_3^{VI}$	$A_8^{II} B^{III} C_{11}^{VII}$	—	$A^{II} B^{IV} C_3^{VI}$	$A_6^{II} B^{IV} C_8^{VII}$	$A_4^{II} B_2^V C_9^{VI}$	$A_4^{II} B^V C_5^{VII}$
$A^{II} B_2^{VII}$; $A^{IV} B_2^{VI}$	5.33	$A^I B_3^V C_8^{VI}$	$A_3^I B^V C_8^{VII}$	$A_4^I B^{VI} C_{10}^{VII}$	—	—	—	$A^{II} B_2^{VII}$	—	$A^{IV} B_2^{VI}$	$A_3^{II} B^{IV} C_{10}^{VII}$	$A_4^{II} B_2^V C_6^{VI}$	$A^{II} B_2^{VII}$
$A_2^V B_5^{VI}$	5.71	$A_2^V B_5^{VI}$	$A_5^I B_3^V C_{20}^{VII}$	$A_7^I B_3^{VI} C_{25}^{VII}$	—	—	—	$A^{II} B^{III} C_5^{VII}$	—	—	$A_3^{II} B^{IV} C_{10}^{VII}$	$A_2^V B_5^{VI}$	$A_5^{II} B^V C_{15}^{VII}$
$A^{III} B_3^{VII}$	6.00	—	$A^I B^V C_6^{VII}$	$A_3^I B_2^{VI} C_{15}^{VII}$	—	—	—	$A^{III} B_3^{VII}$	—	—	$A^{II} B^{IV} C_6^{VII}$	—	$A_2^{II} B^V C_9^{VII}$

TABLE 2 (Continued)

Formula of binary prototype	Electrons per atom	II–VI–VII	III–IV–V	III–IV–VI	III–IV–VII	III–V–VI	III–V–VII	III–VI–VII	IV–V–VI	IV–V–VII	IV–VI–VII	V–VI–VII
$A_3^I B^V$	2.0	—	—	—	—	—	—	—	—	—	—	—
$A_5^{II} B_2^{III}$	2.29	—	—	—	—	—	—	—	—	—	—	—
$A_2^I B^{VI}$; $A_2^{II} B^{IV}$	2.67	—	—	—	—	—	—	—	—	—	—	—
$A_3^{II} B_2^V$	3.20	—	—	—	—	—	—	—	—	—	—	—
$A_4^{III} B_3^V$	3.42	—	$A_4^{III} B_3^{IV}$	$A_4^{III} B_3^{IV}$	$A_4^{III} B_3^{IV}$	—	—	—	—	—	—	—
$A^{III} B^V$; $A^{II} B^{VI}$ $A^I B^{VII}$	4.00	$A^{II} B^{VI}$	$A^{III} B^V$	$A_2^{III} B^{IV} C^{VI}$	$A_3^{III} B_2^{IV} C^{VII}$	$A^{III} B^V$	$A^{III} B^V$	—	—	—	—	—
$A_3^{IV} B_4^V$	4.57	$A_3^{II} B_2^V C_2^{VII}$	$A_3^{IV} B_4^V$	$A_6^{III} B^{IV} C_7^{VI}$	$A_9^{III} B_5^{IV} C_7^{VII}$	$A_3^{III} B^V C_3^{VI}$	$A_6^{III} B_5^V C_3^{VII}$	—	$A_3^{IV} B_4^V$	$A_3^{IV} B_4^V$	—	—
$A_2^{III} B_3^V$	4.80	$A_2^{II} B^{VI} C_2^{VII}$	—	$A_2^{III} B_3^{VI}$	$A_3^{III} B^{IV} C_2^{VII}$	$A_2^{III} B_3^V$	$A_4^{III} B_3^V C_5^{VII}$	—	$A_2^{IV} B_2^V C^{VI}$	$A_4^{IV} B_5^V C^{VII}$	$A^{IV} B_2^V$	—
$A^{II} B_2^{VII}$; $A^{IV} B_2^{VI}$	5.33	$A^{II} B_2^{VII}$	—	$A^{IV} B_2^{VI}$	$A_3^{III} B^{IV} C_5^{VII}$	$A^{III} B^V C_4^{VI}$	$A_2^{III} B^V C_3^{VII}$	$A^{III} B^{VI} C_4^{VII}$	$A^{IV} B_2^V$	$A_4^{IV} B_3^V C_3^{VII}$	—	—
$A_2^V B_5^{VI}$	5.71	$A_7^{II} B^{VI} C_{20}^{VII}$	—	—	$A_6^{III} B^{IV} C_{14}^{VII}$	$A_2^V B_5^{VI}$	$A_4^{III} B^V C_9^{VII}$	$A_2^{III} B^{VI} C_4^{VII}$	—	$A_4^{IV} B^V C_7^{VII}$	$A_4^{IV} B_3^{VI} C_2^{VII}$	$A_2^V B_5^{VI}$
$A^{III} B_3^{VII}$	6.00	$A_3^{II} B^{VI} C_{12}^{VII}$	—	$A^{III} B_3^{VII}$	$A^{III} B_3^{VII}$	—	$A^{III} B_3^{VII}$	$A^{III} B_3^{VII}$	—	$A_2^{IV} B^V C_5^{VII}$	$A^V B^{VI} C_2^{VII}$	$A^V B_2^{VI} C^{VII}$

trons per atom can exhibit semiconducting properties (for example, Li_3Bi, Na_3Sb). Obviously, the lower limit separating semiconductors from metallic alloys lies somewhere close to two electrons per atom. This is also confirmed by an examination of valence combinations of atoms with fewer than two electrons per atom.

Chemical compounds with 1.14 and 1.33 electrons per atom do not exist because elements of the second, and third groups are not typical cations and do not exhibit a tendency to complete their electron shells to the octet level. Compounds with 1.60 electrons per atom are also unknown; an alloy of copper and tin, which can be attributed the formula Cu_4Sn, has the typical metallic crystal structure and no semiconducting properties.

Thus, the limit of the formation of valence compounds coincides with the limit of the formation of alloys exhibiting semiconducting properties.

There is no upper limit of the number of electrons per atom in chemical compounds which exhibit semiconducting properties. This conclusion follows not only from the fact that the element with the highest number of electrons per atom (seven) is a semiconductor, but also from the fact that known binary compounds with a large number of electrons per atom exhibit semiconducting properties even when their structure is molecular (for example, SnI_4). However, the mechanism of electron transport in such crystals, in which the ionic−covalent forces are limited to molecular dimensions, is somewhat different than that in classical semiconductors with the coordination structure. Therefore, we can arbitrarily draw the upper limit at the number of electrons at which substances with molecular bonds begin to form, i.e., at six electrons per atom.

We shall now consider ternary valence compounds. All simple and complex valence chemical compounds are formed in accordance with simple laws which are related to the definite number of electrons per atom. It will be shown in the next chapter that graphical and analytic methods can be used to calculate quite simply the composition of possible valence compounds of any complexity. We shall later return to the problem to what extent such calculations can be used to predict the actual existence of such phases. At this point, we shall consider which ternary compounds can exist in the valence state with a definite number of electrons per atom. We shall examine Table 2, which lists, in increasing order of the number of electrons per atom (within the limits deduced in the preceding paragraphs), those ternary compounds which are electronic analogs of the corresponding valence binary compounds.

We shall consider only the types given in Table 2 because analysis of all valence compounds is outside the scope of our discussion.

The binary prototypes ($A^{III}B^V$, $A^{II}B^{VI}$, and A^IB^{VII}) of ternary four-electron compounds have two types of structure. With few exceptions, the AB compounds in which cations are elements of the B subgroups have the tetrahedral structures of wurtzite and sphalerite, while the compounds of the same type with cations belonging to the A subgroups have the octahedral structure of sodium chloride. Molecular structures are found in compounds of hydrogen [2]. The formation of binary tetrahedral compounds and of their elemental analogs (diamond, silicon, germanium, and gray tin) is due to the appearance of the stable sp^3 hybrid electron configuration. In octahedral phases, the s^2p^6 configuration is also very stable; it is the analog of the structure of the electron shells of the inert gases. The types of ternary compound which are known to exist are framed in heavy lines in Table 2.

We can see from Table 2 that its central part is occupied by ternary four-electron compounds and a considerable number of such compounds is already known. All the known ternary four-electron compounds have semiconducting properties (they are described in detail in the following chapters of the present monograph). These compounds have lattices which are derivatives of the ZnS and NaCl structures.

Typical representatives of compounds in which the anion sublattice consists of atoms of the fifth group and the cation sublattice consists of atoms of the B subgroups have tetrahedral structures. Similar compounds containing atoms of the A subgroups have not been investigated much. According to the available information, some of them have a molecular structure (calcium cyanamide). Representative compounds with elements of the sixth group in the anion sublattice frequently have structures similar to those of ZnS and NaCl (for example, Li_3PS_4 has an octahedral structure and Cu_3AsSe_4 has a tetrahedral structure).

Many nonequiatomic compounds have also been prepared but their properties have not been investigated in sufficient detail. One can distinguish several groups of nonequiatomic ternary compounds whose defect structure is close to that of sphalerite if the cations are atoms of the B subgroups. These groups include compounds of the following types: $A_2^I B^{II} C_4^{VII}$, $A^{II} B_2^{III} C_4^{VI}$, $A_2^{II} B^{IV} C_3^{VI}$, $A^{III} B^V C_3^{VI}$, $A^{II} B^V C_4^{VI}$, $A^{II} B^{IV} C_3^{VI}$. The most thoroughly investigated representatives of the first two types of such ternary compounds are described in Chapter V.

When the cation in such ternary compounds is an atom of the A subgroup or when the anion is oxygen, the structure is usually octahedral. For example, $AlSbO_4$ (5.33 electrons per atom) has an octahedral structure of the rutile type.

This completes our general discussion of the crystallochemical properties of the class of semiconducting compounds.

Physicochemical Criterion for the Formation of Ternary Compounds

In the preceding section, we have defined the position of ternary diamond-like semiconductors among other semiconducting materials obeying the same laws of formation. This division of semiconducting compounds into groups is necessary to interpret physicochemical semiconducting properties on the basis of their variation with composition and structure in a given group, which allows us to predict the properties of substances not yet investigated and, which is more important, to predict, to some extent, the possibility of the formation of new types rather than individual compounds. The qualification "to some extent" is made deliberately. The extensive material presented in this monograph shows that analysis of the energy and volume parameters of atoms forming ternary compounds frequently makes it possible to determine the tendency of substances to crystallize in a given structure (octahedral or tetrahedral). The main properties of the compounds are correlated, as demonstrated in the present monograph, with the composition and therefore predictions can be made but they are not absolutely reliable. In other words, we can draw some preliminary conclusions about the structure and properties of substances once they are obtained in the form of independent chemical compounds.

However, so far it has not been possible to predict the actual formation of ternary compounds.

Analysis of the $T-X$ phase diagrams of binary systems, in which binary tetrahedral phases can be formed, can yield some information on the departures of the stability of the sp^3 electron configuration which result, in the limit, in the disappearance of chemical interaction when the principal quantum numbers of the atoms forming part of the system are increased. In $A^{III}B^V$ systems, the chemical interaction disappears in the transition from antimony to bismuth in the systems $Ga-B^V$ and $Al-B^V$. The phase diagrams of bismuth and gallium and of bismuth and aluminum show a limited solubility of the components in the liquid state, a very low solubility in the solid state, and the absence of compounds.

The nature of the phase diagrams of $In-B^V$ systems changes very markedly when antimony is replaced with bismuth. The chemical interaction does not disappear but a second com-

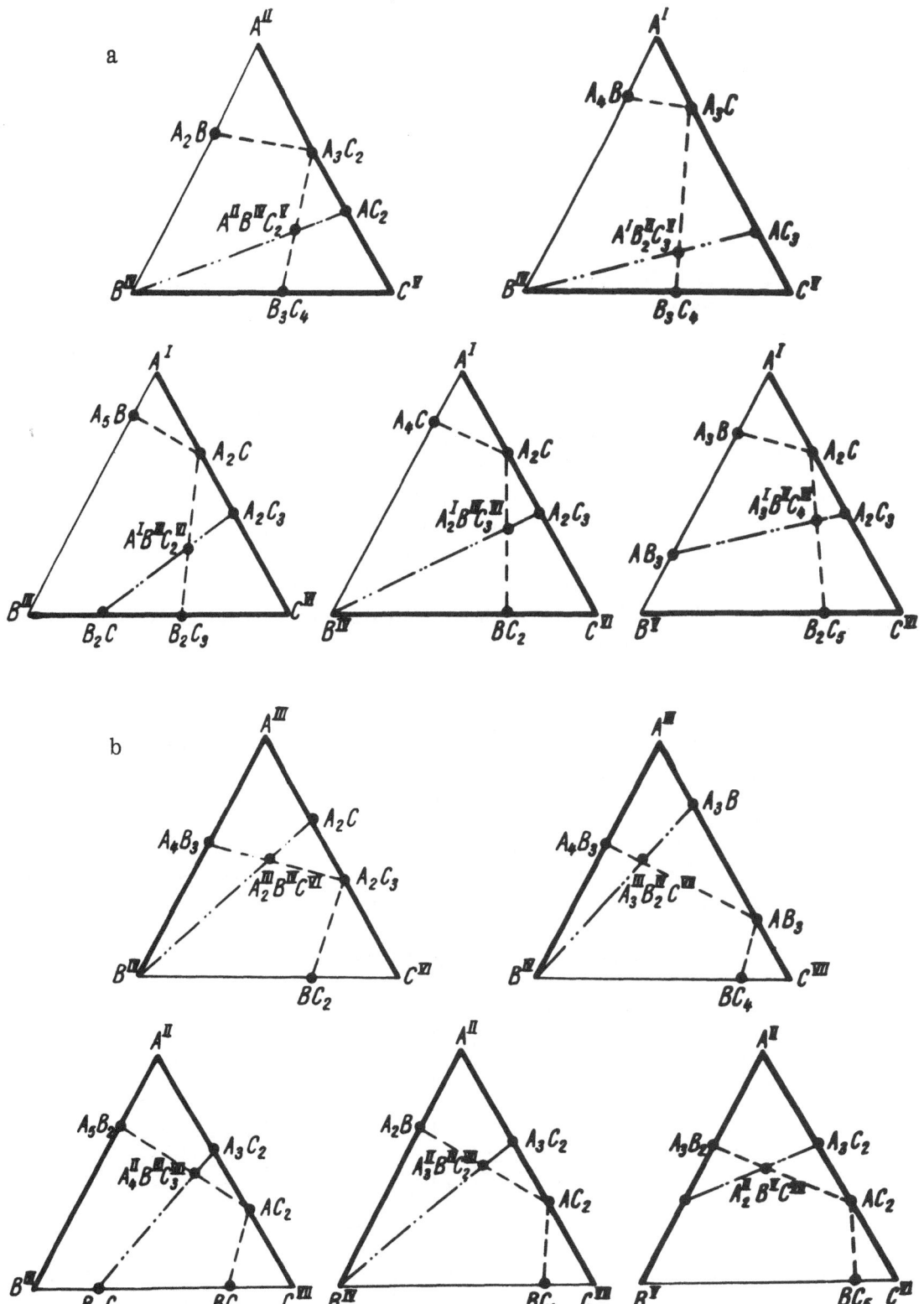

Fig. 1. Formation of ternary two-cation (a) and two-anion (b) valence compounds (the thick lines represent the basic binary systems; the dashed lines represent valence compounds; the chain lines represent compounds with four electrons per atom).

pound appears and the structure of the AB compound changes. Instead of the tetrahedral sphalerite structure with the sp^3 hybridization of electrons (as in compounds of indium with antimony, arsenic, and phosphorus), we obtain a layered tetragonal structure with a more complex type of chemical binding.

Thus, in the case of binary systems, we can see that, although the conditions for the formation of a compound are fulfilled, the expected phase is either not formed at all or does not have the predicted structure.

The phase diagrams of ternary systems in which tetrahedral phases can form have hardly been investigated. Therefore, the possibility of chemical interaction in these systems has to be judged from the phase diagrams of binary systems forming part of a ternary system.

Examination of the Gibbs triangles of A−B−C systems, in which ternary tetrahedral phases can form (Fig. 1), shows that the formation of ternary phases is governed by the chemical interaction in A−C and B−C systems (in the case of two-cation phases) and by the chemical interaction in A−B and A−C systems (in the case of two-anion phases), i.e., in cation−anion systems. We shall call them the basic systems. The nature of the interaction in A−B systems in the former case and in B−C systems in the latter case is obviously not decisive although it may be of some importance. It follows from Fig. 1 that, if there is no chemical interaction in the basic systems, ternary tetrahedral phases (and other stable phases) are not formed. This follows also from the crystallochemical concept of valence ternary tetrahedral compounds, consisting of cations and anions held in the lattice by covalent as well as fairly strong ionic forces. The absence of the ionic interaction between atoms located in the cation and anion sublattices reduces the difference between them and prevents the formation of a ternary compound (but favors the formation of solid solutions).

In spite of the obviousness of the conclusion that ternary phases cannot be formed when there is no chemical interaction in one of the basic systems, this point has until now been ignored in discussions of the possibility of the formation of tetrahedral and octahedral phases.

First, we must specify more exactly what we understand by the chemical interaction. In the formation of ternary compounds, the presence of eutectic phase diagrams and of diagrams typical of solid solutions may be regarded as evidence of the absence of chemical interaction,

TABLE 3. Ternary Two-Cation $A_2^{III}B^{IV}C^{VI}$ Compounds

C^{VI} $\frac{B^{IV}}{A^{III}}$	Oxides					Sulfides					Selenides					Tellurides				
	C	Si	Ge	Sn	Pb	C	Si	Ge	Sn	Pb	C	Si	Ge	Sn	Pb	C	Si	Ge	Sn	Pb
B	X	X	X	≡	≡	X	X	X	≡	≡	?	?	?	≡	≡	?	?	?	≡	≡
Al	+	≡	≡	≡	≡	X	≡	≡	≡	≡	X	≡	≡	≡	≡	X	≡	≡	≡	≡
Ga	?	≡	≡	≡	≡	?	≡	≡	≡	≡	?	≡	≡	≡	≡	?	≡	≡	≡	≡
In	?	≡	≡	≡	≡	?	≡	≡	≡	≡	?	≡	≡	≡	≡	?	≡	≡	≡	≡
Tl	?	≡	≡	≡	≡	?	≡	≡	≡	≡	?	≡	≡	≡	≡	?	≡	≡	≡	≡
Sc	X	?	?	?	?	X	?	?	?	?	X	?	?	?	?	X	?	?	?	?
V	X	X	?	?		X	X	?	?		X	X	X	?	?	X	X	X	?	?
La	X	X	X	X	X	X	X	X	X	X	X	X	X	X	X	X	X	X	X	X

and only the formation of chemical compounds in binary systems shows whether such an interaction exists.

We shall consider some specific ternary systems in order to illustrate the application of the physicochemical criterion used in the prediction of the formation of ternary compounds. We shall begin with single-cation compounds of the $A_2^{III}B^{IV}C^{VI}$ type, whose existence has been disputed for a long time (cf. Chapter V).

$A_2^{III}B^{IV}C^{VI}$ compounds include oxides, sulfides, selenides, tellurides, and polonides. Since there is hardly any information on polonides of elements in the third and fourth groups, they are not included in Table 3.

We shall now consider A−B and A−C binary systems from the point of view of the formation of binary compounds. Table 3 lists the results reported for these systems in [43, 45, 50, 61]; here, A is understood to represent elements belonging to the main or secondary subgroups. In Table 3, as well as in the subsequent three tables, four symbols are used. If a ternary compound exists and its formation is in agreement with the physicochemical criterion, the corresponding square in the table is marked with a simple cross. If one of the basic systems shows no chemical interaction, the square is shown shaded. If there is no information about either or both systems, a question mark is used. If the corresponding compound is unknown, but both basic systems have chemical compounds, the corresponding square is marked with a diagonal cross, which indicates that the formation of such a compound is likely.

Chemical interaction has been found in all the $A^{III}C^{VI}$ systems investigated so far. Thus, the possibility of the existence of ternary compounds of the type listed in Table 3 is governed by the chemical interaction in $A^{III}B^{IV}$ systems, i.e., by the existence of carbides, silicides, germanides, stannides, and plumbides of elements of the third group. Boron does not react with tin or lead. Aluminum forms only carbides. Gallium, indium, and thallium do not form compounds with elements in the fourth group (no information is available on carbides of these elements).

Thus, the possibilities of the formation of $A_2^{III}B^{IV}C^{VI}$ compounds with elements of the third B subgroup are very limited. If we exclude those ternary systems in which one of the basic systems has not yet been investigated, we can only assume that boron compounds may exist. If such compounds were to be found, this would confirm the semi-intuitive prediction that the precondition for the formation of a tetrahedral ternary compound of this type is simply the existence of any compound in one of the basic systems, even if the compound in a basic system does not obey the valence rule, is not of the four-electron type, or is nonequiatomic.

Apart from aluminum oxycarbide, the existence of aluminum sulfo-, seleno-, and tellurocarbides is very likely because the corresponding binary systems include compounds whose nature and structure are close to that of aluminum oxide. None of the lower elements in the third subgroup combines with any of the elements in the fourth group. Therefore, the corresponding ternary compounds do not exist and we can see why, of all the possible two-anion compounds of the $A_2^{III}B^{IV}C^{VI}$ type, only Al_2CO has been prepared.

The situation is completely different in the case of compounds of this type containing elements of the third A subgroup. It is evident that the formation of scandium oxy-, sulfo-, seleno-, and tellurocarbides should be possible.

Chemical interaction is observed in all the investigated yttrium systems and, therefore, yttrium two-anion compounds are possible. All the basic binary systems based on lanthanum have been investigated and in all of them chemical interaction does occur. Consequently, we may assume that two-anion compounds of lanthanum exist.

TABLE 4. Ternary Two-Cation $A^{II}B^{IV}C_2^V$ Compounds

C^V	Nitrides					Phosphides					Arsenides					Antimonides					Bismuthides				
$\frac{B^{IV}}{A^{II}}$	C	Si	Ge	Sn	Pb	C	Si	Ge	Sn	Pb	C	Si	Ge	Sn	Pb	C	Si	Ge	Sn	Pb	C	Si	Ge	Sn	Pb
Be						?						?	?	?					?						
Mg						?																			
Zn						?																			
Cd						?																			
Hg						?	?	?	?																
Ca						?																			
Sr						?																			
Ba						?																			
Ra						?	?	?	?			?	?	?					?						

The question of the influence of a strong chemical interaction in nonbasic systems on the formation of ternary compounds is worth discussing. Obviously, such an interaction does not exclude the formation of ternary compounds, as indicated by aluminum oxycarbide, in which the nonbasic CO system exhibits chemical interaction. There are many similar examples of the formation of ternary compounds in other systems.

$A^{II}B^{IV}C_2^V$ compounds provide extensive material for the analysis of the influence of the nature of the interaction of components in binary systems (Table 4). This group of compounds, which has been discovered relatively recently and is being studied intensively, already includes twenty compounds formed by elements of the tetrahedral (B) and octahedral (A) subgroups.

Two-cation compounds of the $A^{II}B^{IV}C_2^V$ type include nitrides, phosphides, arsenides, antimonides, and bismuthides. The basic systems are $A^{II}C^V$ and $B^{IV}C^V$. Obviously, the formation of ternary bismuthides is not very likely because there is no chemical interaction in any of the $A^{IV}-Bi$ systems.

We cannot expect the formation of compounds in the antimonide group either, because of the absence of the chemical interaction in $B^{IV}-Sb$ systems.

The exception to this rule is represented by compounds of tin because an ordered intermediate phase, β'-SnSb, is formed in the Sn−Sb system.

We may assume that the compounds $ZnSnSb_2$ and $CdSnSb_2$ do exist although, because of the special properties of the tin−antimony system, they melt incongruently and cannot be prepared by standard methods (fusion of elements).

The scarcity of information on the binary phase diagrams of systems based on beryllium, mercury, strontium, and radium prevents us from making any predictions about the formation of the corresponding ternary compounds.

The formation of $MgSnSb_2$, $CaSnSb_2$, and $BaSnSb_2$ compounds is possible but, in all these cases, the influence of the strong chemical interaction in the nonbasic system $A^{II}B^{IV}$, which is absent from systems of elements of the tetrahedral subgroup, is not clear. However, this chemical interaction has not prevented the formation of, for example, $MgSiN_2$, $MgSiP_2$, etc.

In the arsenide group, we must exclude all ternary systems containing the carbon – arsenic and lead – arsenic binary systems, in which no chemical interaction has been observed. As in the case of antimonides, the lack of information on the phase diagrams makes it difficult to predict the formation of beryllium compounds. Zinc and cadmium compounds have been obtained; their basic binary phase diagrams include chemical compounds but there is no interaction in the nonbasic systems. Ternary compounds of magnesium and elements in the octahedral subgroup (with the exception of radium, for which no phase diagrams are available) are, in general, possible, but the strong interaction of elements in the second and fourth groups may distort the structure (when the relationship between the energies of formation of $A^{II}B^{IV}$ compounds, on the one hand, and $A^{II}C^{V}$ and $B^{IV}C^{V}$, on the other, is unfavorable).

In the phosphide subgroup, all the compounds containing lead, mercury, and radium cannot be discussed because of the absence of information on the lead – phosphorus, mercury – phosphorus, and radium – phosphorus systems.

It is difficult to predict the existence of compounds containing carbon because the information on the carbon – phosphorus system is limited. There is only one mention of a compound, C_3P, which is stable at room temperature. If this compound does indeed exist and its energy of formation is comparable with the energies of the formation of the corresponding acetylides and carbides, then ternary compounds of this type (carbophosphides) may exist.

Ternary zinc and cadmium phosphides have been prepared and their basic phase diagrams are similar to the phase diagrams of arsenides. In view of the similarity of the basic (and nonbasic) phase diagrams of beryllium, zinc, and cadmium compounds, we can definitely conclude that ternary compounds of beryllium do exist.

Ternary nitrides also include the well-known cyanamides of elements in the second group (magnesium, zinc, and cadmium). The ternary compounds $BeSiN_2$ and $MgSiN_2$ have also been synthesized. The former has a structure similar to that of wurtzite. The phase diagrams of the basic systems of ternary nitrides of beryllium, magnesium, zinc, and cadmium, on the one hand, and of carbon, silicon, and germanium, on the other, are similar. It is very likely that these nitrides, which have not yet been synthesized (out of a group of 16 possible compounds), will be prepared in the immediate future. The difference between the nonbasic system of magnesium from the other systems of this group is obviously of no importance because the existence of two ternary magnesium nitrides has already been reported. Ternary mercury nitrides of this type are also likely to exist. However, there are some doubts about these nitrides. The doubts arise because chemical interaction is observed in the basic mercury – nitrogen system, but compounds which are formed, including the valence compound Hg_3N_2, are explosive and dissociate easily. This may prevent the formation of ternary compounds. However, the lattice of ternary compounds may be considerably stronger because of a more symmetrical distribution of the sp^3 bonds.

All that we have said about mercury nitrides applies even more strongly to ternary nitrides containing lead. Only the nonvalence compound, lead azide, which is a strong explosive, is formed in the lead – nitrogen system.

In the octahedral subgroup of nitrides, we have, in contrast to all the other octahedral subgroups, two ternary nitrides. One of them is calcium cyanamide, which is a well-known fertilizer. The other is much more "exotic": $CaSiN_2$. Calcium cyanamide has a complex crystal structure in which the CN_2 groups are retained. This complex molecular-type structure may be due to two causes: the presence of light carbon atoms, which favors the formation of a molecular structure; and chemical interaction in the nonbasic system.

TABLE 5. Ternary Two-Cation $A^I B^{III} C_2^{VI}$ Compounds

C^{VI}	Oxides					Sulfides					Selenides					Tellurides				
$\dfrac{B^{III}}{A^I}$	B	Al	Ga	In	Tl	B	Al	Ga	In	Tl	B	Al	Ga	In	Tl	B	Al	Ga	In	Tl
Cu											?									
Ag											?									
Au											?									
Li											?									
Na											?									
K											?									
Rb											?									
Cs											?									

The similarity of all the basic phase diagrams of compounds of the octahedral subgroup and the fact that they obey the physicochemical criterion suggest that all of them, with the exception of lead nitrides, will be prepared in the near future.

We shall now analyze the interaction of components in the binary systems on which ternary compounds of the $A^I B^{III} C_2^{VI}$ type are based (Table 5).

The results presented in Table 5 show that at least 40 compounds of this type are known to exist; this is due to the strong chemical interaction in the majority of the basic systems. The existence of such a large number of compounds is possibly also favored by the complete absence of the chemical interaction in all the nonbasic systems. The situation is not clear in the case of ternary boron selenides because of the lack of information on the boron−selenium system.

Since no stable compounds exist in the boron−tellurium system, it follows that there are no ternary compounds of boron and tellurium.

The other three types of ternary compound, for which typical examples are known, are difficult to analyze because of their complexity. The formula units of these ternary compounds consist, in contrast to those discussed so far, not of four atoms but of six or even eight atoms ($A_2^I B^{IV} C_3^{VI}$, $A^I B_2^{IV} C_3^V$, and $A_3^I B^V C_4^{VI}$). The number of atoms in the formula unit gives some idea about the short-range distribution of atoms. In the ternary compounds considered, we are dealing with unequal numbers of different atoms in the cation sublattice (2:1, 1:2, and 3:1). $A_2^I B^{IV} C_3^{VI}$ compounds (Table 6) have five subgroups: oxides, sulfides, selenides, tellurides, and polonides. The polonide subgroup will not be discussed because the corresponding binary systems have not been investigated.

Predictions about this group of compounds are less reliable but, even so, we can conclude that the formation of ternary compounds of gold, sulfur, and selenium or of ternary compounds containing carbon and tellurium is very unlikely. However, many substances are known to exist in the group under consideration; some of them, for example, sodium carbonate, have been found unexpectedly to be analogs of semiconductors such as Cu_2GeSe_3. A crystallochemical analysis of this group would be very interesting but it is likely to be difficult.

In addition to the tetrahedral compounds formed by elements of the B subgroups, the octahedral subgroup includes complex silicate structures. In contrast to the other groups

TABLE 6. Ternary Two-Cation $A_2^I B^{IV} C_3^{VI}$ Compounds

C^{VI}	Oxides					Sulfides					Selenides					Tellurides				
$\dfrac{B^{IV}}{A^I}$	C	Si	Ge	Su	Pb	C	Si	Ge	Su	Pb	C	Si	Ge	Sn	Pb	C	Si	Ge	Su	Pb
Cu																				
Ag																				
Au																				
Li																				
Na																				
K																				
Pb																				
Cs																				

which we have discussed, where the physicochemical criterion ensures, in most cases, success in the synthesis of a semiconducting compound, departures from this criterion are often observed in the group now being considered.

Ternary compounds of other types, listed in Table 2, have not yet been obtained, although all of them satisfy the physicochemical criterion.

We must now consider the problem of the likelihood of the participation of the d elements of the fourth group in compounds of the types considered so far, bearing in mind the presence of elements of the B and A subgroups.

In the first group of compounds which we shall consider $(A_2^{III} B^{IV} C^{VI})$, an element of the fourth group forms part of the anion sublattice. The electron structure of titanium and its analogs prevents their participation in the form of anions.

The following remarks can be made about the participation of titanium and its analogs in $A^{II} B^{IV} C_2^V$ compounds. Obviously, the participation of titanium in compounds of this type is either unlikely or, at most, probable only when combined with elements of the second A subgroup of the periodic system.

The participation of scandium, yttrium, lanthanum, and lanthanides in the $A^I B^{III} C_2^{VI}$ group is likely in all cases. Some of these compounds, $LiScO_2$, $NaScO_2$, $LiVO_2$, and others have already been obtained [70]. In these cases, the formation of structures similar to NaCl is very likely.

The remarks made about the participation of titanium and its analogs in compounds of the $A^{II} B^{IV} C_3^V$ type apply also to compounds of the $A_2^I B^{IV} C_3^{VI}$ type. It is also known that titanium is present in a compound of this type belonging to the octahedral subgroup (Li_2TiO_3) [32].

We shall now summarize our discussion of all the possible valence ternary compounds, including diamond-like semiconductors, which are the subject of the present monograph.

The physicochemical criterion has been selected because compositions of some types of ternary valence compound have been calculated by mathematical methods (cf. Chapter II).

In these methods, we express mathematically two requirements which govern membership of the class of valence compounds and of subgroups in this class (the conditions of valence and number of electrons per atom). These two requirements or conditions are necessary but insufficient for the formation of ternary compounds.

The insufficiency of these conditions is self-evident: they do not take into account the metallization of chemical bonds and the changes in the properties of elements of the same group in the periodic system with increasing atomic mass. The application of the physicochemical criterion to four-electron valence phases makes it possible to exclude a very large number of combinations, which narrows down the search for new compounds.

Analysis of the possibility of the existence of ternary compounds on the basis of the physicochemical criterion indicates that all the synthesized and existing compounds satisfy the conditions of the interaction of atoms in binary cation—anion systems. Not a single exception to this rule is known.

None of the known methods of predicting the existence of ternary compounds gives absolute accuracy. The physicochemical criterion can be used to predict the possibility of the formation of ternary compounds of any coordination. The mathematical criterion makes it possible to predict the existence of tetrahedral coordination phases. Both criteria must be satisfied to obtain a ternary tetrahedral compound.

However, even in those cases when the formation of a compound is possible, its actual existence is governed by at least another factor. This factor is the relative energy stability of various electron configurations.

One of the most stable electron configurations, which has the minimum free energy, is the sp^3 configuration. However, its energy stability varies with the composition of the compound. It decreases with increasing atomic mass, i.e., when the bonds become metallic (in the case of semiconducting compounds, it is hardly correct to speak of an increase of the statistical weight of delocalized electrons, which is sometimes used to explain the metallization of bonds in semiconductors).

The stability of other electron configurations may increase when the composition is altered. For example, it is known that the p-type nature of the bonding eigenfunctions is stronger for heavier atoms [14]. Thus, the interaction of components of a ternary system may involve competition from another electron configuration, whose stability increases so much as to make it the most likely configuration. This applies to $A^I - B^{IV} - C^{VI}$ systems when the replacement of tin with lead produces not the ternary phase Cu_2PbTe_3 (cf. Table 6), but the stable phase $PbTe$.

Thus, a more accurate prediction of the existence of ternary compounds requires analysis of the free energy of formation of binary systems but, unfortunately, there are very few data on these energies. We can use the physicochemical criterion for ternary four-electron valence compounds, which, like the mathematical criterion, is a necessary but insufficient condition for the formation of ternary phases.

Reverting to the problem of semiconducting properties of the substances considered so far, we may conclude that the classification of chemical compounds also yields a classification of semiconductors.

This conclusion applies fully to the central group of the system of chemical compounds, i.e., to four-electron valence compounds [326] but there are grounds for assuming that, in many other cases, the concept "semiconducting phase" is synonymous with the concept "chemical compound." In general, this means that new ideas in inorganic chemistry can be used in the search for new semiconducting materials and, although intuition still has its place, searches and investigations can and, indeed, are becoming more systematic.

CHAPTER II

FORMATION OF TETRAHEDRAL PHASES

Structure of Semiconducting Compounds.

Tetrahedral Coordination

The properties of semiconducting compounds are governed primarily by the short-range order, i.e., by the chemical nature of atoms, the geometry of their distribution, and the absolute distances between them. The crystal structure of semiconducting compounds is a very important characteristic of the short-range order and, in the majority of such compounds, it can be deduced from the close packing of anions when the tetrahedral and octahedral vacancies are occupied by cations.

The spatial distribution of atoms is governed by the directional nature of the chemical bonds. Tetrahedral bonds produce tetrahedral structures, which, depending on the chemical nature of the atoms forming these structures, may be normal, defect, or "filled" [3]. The majority of compounds with the tetrahedral coordination exhibits the sp^3 hybridization, but there have been reports of the existence of compounds with the tetrahedral coordination with different types of hybridization.

In the case of the sp^3 hybridization, the electrons are distributed between four sp^3 hybrid orbitals. The wave functions of the sp^3 orbitals can be found from four linear combinations of the s and p wave functions [1-3]:

$$\psi_{t_{111}} = \frac{1}{2}(\psi_s + \psi_{p_x} + \psi_{p_y} + \psi_{p_z}),$$
$$\psi_{t_{1\bar{1}\bar{1}}} = \frac{1}{2}(\psi_s + \psi_{p_x} - \psi_{p_y} - \psi_{p_z}),$$
$$\psi_{t_{\bar{1}1\bar{1}}} = \frac{1}{2}(\psi_s - \psi_{p_x} + \psi_{p_y} - \psi_{p_z}),$$
$$\psi_{t_{\bar{1}\bar{1}1}} = \frac{1}{2}(\psi_s - \psi_{p_x} - \psi_{p_y} + \psi_{p_z}).$$

The sp^3 hybrid functions, with probability distributions in the form of asymmetric dumbbells, are directional and form tetrahedral angles of 109.4° with each other. These directions represent the lines joining the center of a regular tetrahedron to its vertices. Figure 2 shows nonhybrid and hybrid orbitals of the s and p electrons.

Quantum-mechanical calculations show that the strength of a bond is governed by the degree of overlap of the orbitals of each pair of electrons participating in that bond. The greater degree of overlap of the orbitals in the sp^3 hybrid bonds, compared with the overlap of the s–p orbitals, results in a lowering of the binding energy. Figure 3 shows graphically the energy decrease in the formation of a compound from isolated atoms: the energy of electrons in the valence band of the compound is lower than the s and p energy levels of isolated atoms [3].

The lower energy and the consequent increase of the strength of the sp^3 hybrid bonds provide the basic energy for the formation of tetrahedral structures.

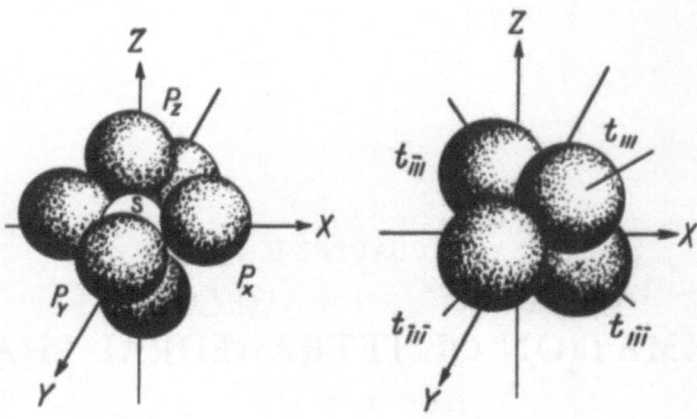

Fig. 2. Nonhybrid and hybrid orbitals
of the s and p electrons.

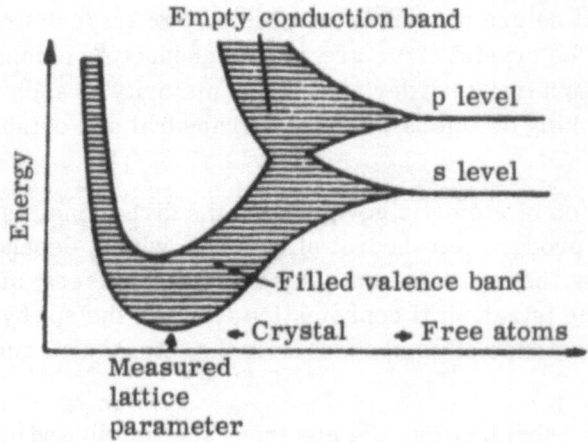

Fig. 3. Schematic representation of the energy
bands for a tetrahedral structure.

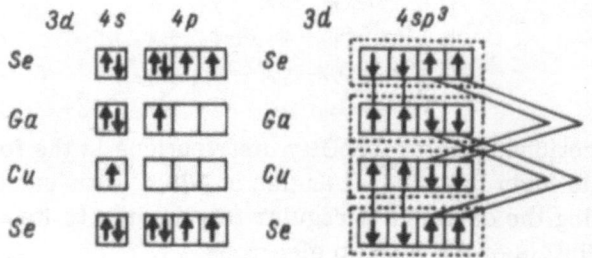

Fig. 4. Formal diagram representing electron
states in the chalcopyrite structure.

The valence bond scheme, based on the sp^3 hybridization, applies not only to binary but also to ternary and more complex compounds. Figure 4 shows formally the diagram of electron states of ternary compounds with the chalcopyrite structure [94]. The continuous lines represent the sp^3 hybrid bonds. However, Parthé [3] has pointed out that such schemes are simplified because, in the chalcopyrite structure (for example, that of $CuGaSe_2$), each group

VI atom has two bonds with two group I atoms and two group III atoms. Moreover, the nonuniform environment distorts the tetrahedron so that the tetrahedral angle departs from 109.4°. In this case, the wave functions of the hybrid orbitals have different admixtures of the s wave function.

In some cases, the formation of mixed p^3 and sp^3 bonds is possible. Thus, for example, $CuSbS_2$ and $CuBiS_2$ have mixed bonds such that A^I and S_{II} atoms have the fourfold coordination, while B^V and S_I atoms have only three nearest neighbors [5].

When electron pairs are not hybridized at all and are in one s and three p states, the eigenfunctions of the p electrons have large positive values along the coordinate axes and large negative values in the opposite directions, and they are directed along three mutually perpendicular axes. This gives rise to the octahedral (and not the tetrahedral) space symmetry.

Group theory is used in [6] to deduce, in the general form, the combinations of the s, p, d, and f orbitals resulting in a given spatial symmetry, including the tetrahedral symmetry.

The chemical nature of atoms governs, in all probability, the nature of the chemical bonds between them, as well as the crystal structure. It follows that there is a close relationship between many properties of compounds and the positions of the component elements in the periodic table.

The nature of chemical bonds can be represented by means of various parameters. These include the ionization potentials, the difference between the electronegativities, and the electron affinity constants.

It has been demonstrated [7, 8] that when the ionization potential of a cation in binary compounds is increased, the nature of the bonds becomes mainly covalent. The ionization potential increases as the element approaches the inert-gas electron structure, and when the charge of the element increases.

The difference between the electronegativities can also be used to estimate the degree of ionicity of the bonds. According to Pauling [1], the degree of the ionicity of bonds can be described by the formula $1 - e^{1/4(x_A - x_B)^2}$, where x_A and x_B are the electronegativities of elements A and B.

An intermediate type of bond can be represented by a wave function $a\psi_{A:B} + b\psi_{A^+B^-}$, which represents a combination of the covalent and ionic bond functions.

In the majority of semiconducting compounds, the difference between the electronegativities Δx is close to unity. Consequently, the ionic component of binding does not exceed 25% [9, 10].

Folberth [13] considered $A^{II}B^{IV}C_2^V$ compounds from the point of view of the electronegativity and concluded that the likelihood of the formation of the chalcopyrite structure increases, from the energy point of view, with the increase of the electronegativity of atoms of the second group relative to atoms of the fourth group. Since the elements in subgroups IB and IIB have relatively large electronegativities, these elements form the well-known compounds with the chalcopyrite structure. This explains the absence of compounds with the antichalcopyrite structure, when the electronegativities of elements in the subgroups VIB and VIIB are considered. These points will be discussed in greater detail in Chapter IV.

If the difference between the electronegativities is used as a measure of the bond ionicity, the principal quantum number of the valence shell of an atom can be regarded as a measure of the directionality of the bonds. When the average principal quantum number $\bar{n} = \Sigma c_i n_i / \Sigma c_i$ (where n_i is the principal quantum number of an atom of i-th type and c_i is the number of such atoms) is increased, the directionality of the electron orbitals decreases and the metallization of bonds is observed.

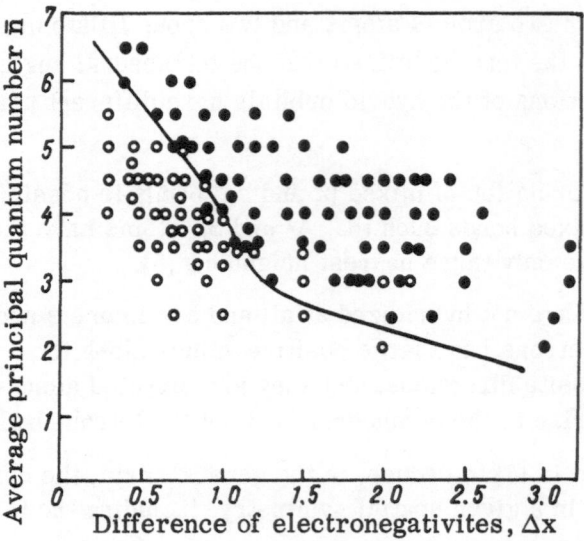

Fig. 5. Normal valence compounds with the zinc-blende, wurtzite, chalcopyrite, and rocksalt structures: ○) tetrahedral coordination, B3, B4, E1₁; ●) octahedral coordination, B1.

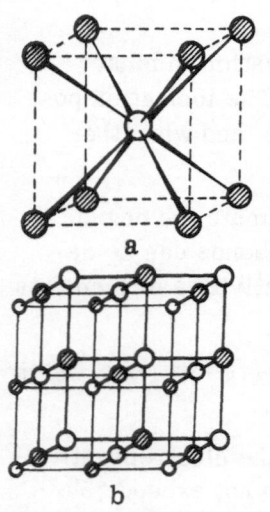

Fig. 6. Structures of: a) cesium chloride; b) rocksalt.

Using $\bar{n}$ and Δx, Pearson [11] has obtained a clear demarcation between the tetrahedral structures of sphalerite, wurtzite, and chalcopyrite, and the octahedral structures (for example, that of rocksalt). Figure 5 shows the demarcation between such structures as a function of the average value of the principal quantum number and the difference between the electronegativities.

The strong metallization of bonds in compounds of heavy metals is responsible for the absence of the chalcopyrite structure in antimony, bismuth, and lead, which all have tetrahedral directional bonds.

Many compounds with typical ionic binding, in which the anion electrons do not have a tendency to occupy the cation quantum states, crystallize in the cubic structure of the CsCl type. When the anion electrons are in the s and p states and do not form the sp^3 hybrids, we obtain the NaCl-type structure.

The zinc-blende structure is formed only when electrons are in the sp^3 hybrid state. The structures of cesium chloride, rocksalt, and sphalerite are shown in Figs. 6 and 9.

Goryunova [12] used the model of an electron lattice, depending on the cores of atoms located at the crystal lattice sites, to show that the tetrahedral coordination in AB-type compounds is formed when the specific electron affinity constant of each of the elements $N' = N/n$ is greater than 7.25 V. This approach to the problem of coordination shows that the tetrahedral and octahedral structures can be divided by a straight line $N'_2 = 6 + 0.1 N'_1$ in the coordinate plane $N'_1 - N'_2$. This dividing line is shown in Fig. 7: the tetrahedral structures lie above the line and the octahedral lattices below it. According to Goryunova, this rule can be extended to more complex compounds.

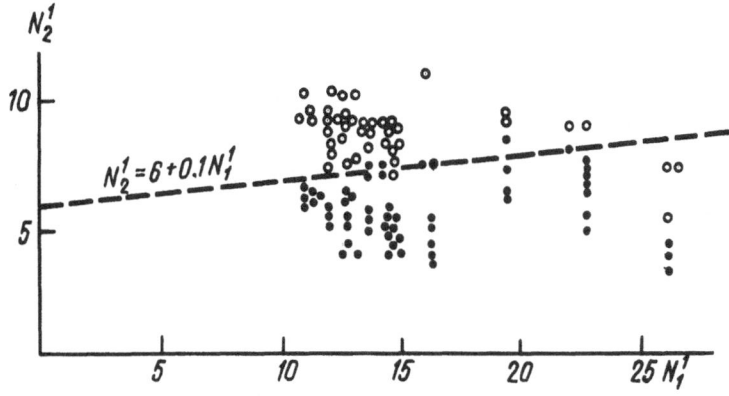

Fig. 7. Demarcation of the structures of AB-type compounds:
O) tetrahedral coordination; ●) octahedral coordination.

Electron Shells of Elements and Crystal Structure

Not all elements can form compounds with the tetrahedral coordination of atoms. Those that do form such compounds are mainly elements of the second and third periods, elements of the B subgroups of the fourth and fifth periods, and some elements of the B subgroup of the sixth period. Table 7, which is based on the results reported in [3, 12], lists the elements which are usually encountered in structures with the tetrahedral coordination.

Elements of the second period, in which the L shell is being filled, participate in the formation of tetrahedral structures. This applies also to elements of the third period (with the exception of sodium), in which the 3s and 3p states are being filled. Further filling of the M shell by the group of the 3d states (in the presence of 4s states) is observed also in elements of the fourth period. Parthé [3] concludes that the participation of such elements as vanadium, manganese, and iron in some compounds with the tetrahedral coordination should be regarded as an exception. However, in our opinion, it can be shown that the participation of transition elements in ternary compounds with the tetrahedral structure is a natural consequence of the application of rules governing the formation of tetrahedral phases.

The presence of the s and d levels with similar energies results in the formation of the sd^3 hybrids. The eigenfunctions of the sd^{3+} hybrids are directed along the space diagonals of a cube, in the positive and negative directions, which correspond to the structure of fluorite (CaF_2), in which each half of the total number of fluorine atoms forms a diamond-like structure with calcium atoms [14].

TABLE 7. Elements forming Tetrahedral Structures

Principal quantum number	Number of valence electrons						
	1	2	3	4	5	6	7
2	Li	Be	B	C	N	O	F
3		Mg	Al	Si	P	S	Cl
4	Cu	Zn	Ga	Ge	As	Se	Br
5	Ag	Cd	In	Sn	Sb	Te	I
6		Hg	Tl		Bi	Po	

The presence of electrons in a d orbital, whose principal quantum number is smaller by unity than that of the valence shell (which is not yet filled completely), exerts a considerable influence on the formation of bonds and the type of structure. The energies of the 3d, 4s, and 4p states of elements in the first transition group (the iron group) are similar. This makes possible the spd hybridization and the sp^2d hybrids give rise to bonds lying in one plane at angles of 90° to each other [1].

Six hybrid functions of the sp^3d^2 hybrid state give rise to the octahedral distribution of bonds. It is known [2] that the s, p, and d functions cannot contribute more than six identical valences.

In some complex compounds, the valence electrons may be in the sp^3 and sp^3d^2 hybrid states. Thus, for example, in $ZnCo_2O_4$, which has the spinel structure, the zinc and oxygen electrons form the sp^3 hybrids and the cobalt electrons form the sp^3d^2 states [14].

The M shell is filled in nickel and then, beginning from copper, the 4s and 4p states are filled gradually. It follows from Table 7 that all the elements in the B subgroups of the fourth period can participate in the formation of tetrahedral structures. Of the elements in the A subgroups of the fifth period, with unfilled 4d shells, only molybdenum can participate in the formation of tetrahedral structures. The B subgroups of the fifth period consist of elements with filled 4d shells, which can take part in the formation of tetrahedral structures, but in the A subgroups of the sixth period only tungsten is known to form such structures.

The heavy elements of the B subgroups of the sixth period do not always form the sp^3 hybrid bonds. No compounds of lead or gold with the tetrahedral coordination are known. It is very likely that this is due to the metallization of bonds in compounds of heavy elements. The energies of the levels 5d, 5f, 6s, and 6p are similar and, therefore, compounds containing elements with these levels can exhibit the $6sp^3$ as well as other types of hybridization.

Generally speaking, we can hardly expect the existence of a large number of tetrahedral compounds with predominantly covalent bonds, in which the f electrons participate. It is assumed in [15] that the tetrahedral sf^3 bonds are not favored by energy considerations and the f electrons tend to form stronger quadratic hybrid sdf^2 orbitals.

Conditions and Criteria for the Formation of Tetrahedral Phases

In the preceding sections of this chapter, we have discussed the conditions for the formation of compounds with the tetrahedral coordination. However, most of these conditions have been deduced from known observations and are of little help in predicting when the tetrahedral coordination will occur. The question of the limits of the existence of tetrahedral phases is very important, as well as the problem of the quantitative criteria which would make it possible to predict the tetrahedral coordination of atoms in the lattice of a given compound. The solution of this problem would make it possible to suggest directions in which to search for new semi-conductors with diamond-like structures.

According to the Magnus rule, when we consider the structure of sphalerite as a close-packed cubic lattice of atoms of one kind with tetrahedral vacancies half-filled with atoms of another kind, we find that the tetrahedral coordination is possible if the ratio of the cation and anion radii lies between the limits 0.22-0.41 and 2.41-4.45.

However, the application of the Magnus rule to compounds with predominantly covalent binding is not fully justified because in such compounds (depending on the ratio of the ionic and covalent components of the binding forces) the values of the ionic radii of the same atoms may be different under different conditions and the use of tabulated radii, obtained for particular compounds, is not permissible. In each of such cases, we have to investigate the distribution of the electron density in the lattice in order to estimate the ionic radii and to determine whether a given assembly of atoms obeys the Magnus rule.

For these reasons, it is not surprising that among known binary compounds with the sphalerite and wurtzite structures, which are analogs of the elemental semiconductors of the fourth group, only about 40% satisfy the Magnus rule. The application of this rule to ternary compounds presents even greater difficulties because of the presence of atoms of two different sizes in the cation (or anion) part of the lattice.

It follows that the Magnus rule is not a helpful criterion for the prediction of the tetrahedral coordination in diamond-like semiconductors.

Grimm and Sommerfeld [16] suggest, on the basis of the experimental results, the following rules for the formation of binary nondefect compounds with the tetrahedral coordination, which are crystallographic analogs of the elemental semiconductors of the fourth group in the periodic table: (1) the components of a compound should belong to groups equidistant from the fourth group; and (2) the average number of valence electrons per atom in the compound should be four.

The majority of the known binary compounds with sphalerite (or wurtzite) structure does satisfy these rules but there are exceptions (manganese sulfide and selenide, as well as compounds of elements of lower periods in the Mendeleev table), which will be discussed later.

It is clear that the Grimm and Sommerfeld rules are inapplicable to more complex compounds. Employing cross substitution, Goodman [10] predicted and synthesized a series of ternary semiconducting phases with diamond-like structures.

Goryunova [12, 18] analyzed in detail the available experimental data and solved the problem of the tetrahedral coordination in compounds with any number of components. Her solution can be formulated as follows: the tetrahedral configuration of atoms in a nondefect structure (diamond, sphalerite, chalcopyrite, wurtzite, enargite, and similar structures) of a compound with an arbitrary number of atoms is possible if (1) the average number of valence electrons per atom of the compound is equal to four; and (2) the valence of each of the components is equal to the number of the group in the periodic table to which this component belongs.

We can easily see that Goryunova's rules include, as special cases, the Grimm—Sommerfeld rules and Goodman's cross substitution idea.

Goryunova's rules for a compound containing A components, n of which are cations, may be written in the following form

$$\sum_{i=1}^{A} B_i x_i = 4,$$

$$\sum_{i=1}^{n} B_i x_i = \sum_{i=n+1}^{A} (8 - B_i) x_i,$$

$$\sum_{i=1}^{A} x_i = 1,$$

where B_i is the number of the group in the periodic system to which the i-th component of the compound belongs, and x_i is the concentration of that component.

Following Goryunova [12], we shall apply these rules to determine all the theoretically possible types of ternary single-anion and single-cation compounds with the tetrahedral coordination.

The formula of a single-anion compound can be written in the form $A_{1-x-y}B_xC_y$, where x and y are the atomic concentrations of the components in the compounds, and A, B, C represent groups of the periodic system to which these components belong.

The first rule of Goryunova can be expressed mathematically in the form

$$A(1-x-y)+Bx+Cy=4,$$ (1)

and the second is given by

$$A(1-x-y)+Bx=(8-C)y.$$ (2)

Solving simultaneously Eqs. (1) and (2) we obtain

$$y=1/2$$ (3)

and

$$x=\frac{8-(A+C)}{2(B-A)}.$$ (4)

For three components, the following conditions must be satisfied:

$$x+y<1 \text{ and } x>0,$$ (5)

and then the following inequalities are obtained from Eqs. (3), (4), and (5):

$$\frac{8-(A+C)}{2(B-A)}>0$$ (6)

and

$$\frac{8-(A+C)}{2(B-A)}<^1/_2.$$ (7)

Goryunova [12] has assumed that A < B < C, which shows that five types of ternary single-cation compound can exist. However, it is well known that transition elements can behave only as cations in compounds, irrespective of their position in the periodic system. Hence, we can replace the condition A < B < C by three other conditions:

$$C-A>0,$$ (8)

$$B-A>0$$ (9)

and

$$B\gtrsim C.$$ (10)

Elementary analysis of the inequalities (6)–(10) shows that 25 types of compound, satisfying Goryunova's rules, can exist.

Table 8 gives the results of our analysis. The first eight types of compound are formal ternary analogs of $A^{III}B^V$ compounds; the next five types are analogs of $A^{II}B^{VI}$ compounds; the remainder are analogs of $A^{IV}B^{IV}$ compounds (for example, silicon carbide). In this context, we can regard elements such as ruthenium or osmium as belonging to the eighth group because they are transition metals and, in some compounds, they exhibit an eightfold valence.

TABLE 8. Types of Ternary Single-Anion Compounds*

No.	A	B	C	x	Type	No.	A	B	C	x	Type
1	1	4	5	1/3	$A^I B_2^{IV} C_3^V$	14	1	5	4	3/8	$A^I B_3^V C_4^{IV}$
2	1	5	5	1/4	$A^I B^V C_2^V$	15	1	6	4	3/10	$A_2^I B_3^{VI} C_5^{IV}$
3	1	6	5	1/5	$A_3^I B_2^{VI} C_5^V$	16	1	7	4	1/4	$A^I B^{VII} C_2^{IV}$
4	1	7	5	1/6	$A_2^I B^{VII} C_3^V$	17	1	8	4	3/14	$A_4^I B_3^{VIII} C_7^{IV}$
5	2	4	5	1/4	$A^{II} B^{IV} C_2^V$	18	2	5	4	1/3	$A^{II} B_2^V C_3^{IV}$
6	2	5	5	1/6	$A_2^{II} B^V C_3^V$	19	2	6	4	1/4	$A^{II} B^{VI} C_2^{IV}$
7	2	6	5	1/8	$A_3^{II} B^{VI} C_4^V$	20	2	7	4	1/5	$A_3^{II} B_2^{VII} C_5^{IV}$
8	2	7	5	1/10	$A_4^{II} B^{VII} C_5^V$	21	2	8	4	1/6	$A_3^{II} B^{VIII} C_3^{IV}$
9	1	3	6	1/4	$A^I B^{III} C_2^{VI}$	22	3	5	4	1/4	$A^{III} B^V C_2^{IV}$
10	1	4	6	1/6	$A_2^I B^{IV} C_3^{VI}$	23	3	6	4	1/6	$A_2^{III} B^{VI} C_3^{IV}$
11	1	5	6	1/8	$A_3^I B^V C_4^{VI}$	24	3	7	4	1/8	$A_3^{III} B^{VIII} C_4^{IV}$
12	1	6	6	1/10	$A_4^I B^{VI} C_5^{VI}$	25	3	8	4	1/10	$A_4^{III} B^{VIII} C_5^{IV}$
13	1	7	6	1/12	$A_5^I B^{VII} C_6^{VI}$						

*Component B in cases B ≥ C represents transition elements of A subgroups.

TABLE 9. Types of Single-Cation Phases

No.	A	B	C	y	Type
1	2	3	7	1/8	$A_4^{II} B^{III} C_3^{VII}$
2	2	4	7	1/6	$A_3^{II} B^{IV} C_2^{VII}$
3	2	5	7	1/4	$A_2^{II} B^V C^{VII}$
4	3	4	6	1/4	$A_2^{III} B^{IV} C^{VI}$
5	3	4	7	1/3	$A_3^{III} B_2^{IV} C^{VII}$

It is known [12] that compounds belonging to types Nos. 1, 5, 9, 10, and 11 (Table 8) have already been discovered and are being investigated extensively (cf. Chapter IV). Compounds of other types have not yet been investigated thoroughly but there are published reports of the existence of some of them (cf. Chapter IV).

We shall now consider ternary single-cation compounds. The formula of such a compound can be written in the form $A_x B_y C_{1-x-y}$. Goryunova's conditions now become

$$Ax + By + C(1 - x - y) = 4 \qquad (11)$$

and

$$(8 - A)x = By + C(1 - x - y). \qquad (12)$$

The simultaneous solution of Eqs. (11) and (12) yields

$$x = 1/2 \qquad (13)$$

and

$$y = \frac{A + C - 8}{2(C - B)}. \qquad (14)$$

In this case, the conditions specified by Eq. (5) can be rewritten as follows:

$$\left. \begin{array}{l} x + y < 1 \\ y > 0. \end{array} \right\} \qquad (15)$$

Bearing in mind the conditions (13) and (15), we obtain two inequalities from Eq. (14):

$$A + C > 8 \tag{16}$$

and

$$A + B < 8. \tag{17}$$

Assuming that $A < B < C$, we obtain:

$$C - A > 0, \tag{18}$$

$$C - B > 0, \tag{19}$$

$$B - A > 0. \tag{20}$$

It follows from the inequalities (16) and (18) that $C > 4$; and from the inequalities (17) and (20), it follows that $A < 4$. We can see from the inequality (16) that the value $A = 1$ must be excluded because C (C is regarded as an anion) cannot be equal to eight (cf. the footnote to Table 8). Thus, $A = 2, 3$. For $A = 2$, it follows from Eq. (17) that $B < 6$, i.e., when Eq. (20) is taken into account, we find that $B = 3, 4$, or 5, while the condition (16) yields only one value: $C = 7$. For $A = 3$, it follows from Eqs. (17) and (20) that $B = 4$, and from Eq. (16) it follows that $C = 6$ or 7.

There are thus five types of ternary single-cation compound which satisfy the tetrahedral coordination rules. Table 9 lists these five types of ternary single-cation compound, the first three of which can be regarded as ternary analogs of $A^{II}B^{VI}$ compounds and the other two as analogs of $A^{III}B^{V}$ compounds.

However, we must point out that Goryunova's rules do not deal with all the aspects of the existence of tetrahedral phases. For example, it is known that attempts to synthesize $A^{II}B^{IV}C_2^{V}$ compounds containing antimony have been unsuccessful. This shows that, in addition to Goryunova's rules, we must take into account other factors, such as the electron configuration, ionic radii, characteristic features of chemical binding, atomic weights of the components, etc.; in other words, the problem of the criteria of the existence of tetrahedral phases cannot yet be regarded as fully solved in spite of the considerable successes achieved so far.

To illustrate the effectiveness of Goryunova's criteria of the existence of ternary tetrahedral phases, let us consider the causes of the appearance of binary defect phases with tetrahedral coordination, for example, phases of the $A_2^{III}N_3^{VI}$ type.

We can show that these phases also satisfy Goryunova's rules if they are regarded as ternary compounds in which atoms of the third component are defects (whose valence is equal to zero).

The formula of such a defect compound can be written in the form $A_{1-x-y} B_x C_y$ and we then assume that A is equal to zero. Instead of the inequalities (6) and (7), we then obtain

$$\frac{8 - C}{2B} > 0$$

and

$$\frac{8 - C}{2B} < {}^1/_2,$$

and hence

$$8 - C > 0 \tag{21}$$

TABLE 10. Types of Binary Tetrahedral Defect Phases

No.	B	C	x	Type	Typical compounds	No.	B	C	x	Type	Typical compounds
1	2	7	1/4	$B^{II} C_2^{VII}$	BeF_2	15	5	5	3/10	$B_3^V C_5^V$	—
2	3	7	1/6	$B^{III} C_3^{VII}$	$AlCl_3$	16	6	5	1/4	$B^{VI} C_2^V$	—
3	4	7	1/8	$B^{IV} C_4^{VII}$	SiI_4, SiF_4	17	7	5	3/14	$B_3^{VII} C_7^V$	—
4	5	7	1/10	$B^V C_5^{VII}$	PCl_5, PBr_5	18	8	5	3/16	$B_3^{VIII} C_8^V$	—
5	6	7	1/12	$B^{VI} C_6^{VII}$	—	19	5	4	2/5	$B_4^V C_5^{IV}$	—
6	7	7	1/14	$B^{VII} C_7^{VII}$	—	20	6	4	1/3	$B_2^{VI} C_3^{IV}$	—
7	8	7	1/16	$B^{VIII} C_8^{VII}$	—	21	7	4	2/7	$B_4^{VII} C_7^{IV}$	—
8	3	6	1/3	$B_2^{III} C_3^{VI}$	Ga_2Se_3	22	8	4	1/4	$B^{VIII} C_2^{IV}$	—
9	4	6	1/4	$B^{IV} C_2^{VI}$	SiS_2, GeS_2	23	6	3	5/12	$B_5^{VI} C_6^{III}$	—
10	5	6	1/5	$B_2^V C_5^{VI}$	P_2O_5	24	7	3	5/14	$B_5^{VII} C_7^{III}$	—
11	6	6	1/6	$B^{VI} C_3^{VI}$	—	25	8	3	5/16	$B_5^{VIII} C_8^{III}$	—
12	7	6	1/7	$B_2^{VII} C_7^{VI}$	—	26	7	2	3/7	$B_6^{VII} C_7^{II}$	—
13	8	6	1/8	$B^{VIII} C_4^{VI}$	—	27	8	2	3/8	$B_3^{VIII} C_4^{II}$	—
14	4	5	3/8	$B_3^{IV} C_4^V$	Ge_3N_4, Si_3N_4	28	8	1	7/16	$B_7^{VIII} C_8^I$	—

TABLE 11. Simple Phases Obtained from Decomposition of Ternary Defect Phases

Groups of elements	Tetrahedral phase	Binary defect phase	Groups of elements	Tetrahedral phase	Binary defect phase
I—II—VII	$A^I C^{VII}$	$B^{II} C_2^{VII}$	II—V—VI	$A^{II} C^{VI}$	$B_2^V C_5^{VI}$
I—III—VI	$A^I B^{III} C_2^{VI}$	$B_2^{III} C_3^{VI}$	II—V—VII	$A_2^{II} B^V C^{VII}$	$A^{II} C_2^{VII}$
I—III—VII	$A^I C^{VII}$	$B^{III} C_3^{VII}$		$(B^V C_5^{VII})$	
I—IV—VI	$A_2^I B^{IV} C_3^{VI}$	$B^{IV} C_2^{VI}$	II—VI—VII	$A^{II} B^{VI}$	$A^{II} C^{VII}$
I—IV—VII	$A^I C^{VII}$	$B^{IV} C_4^{VII}$		$(B^{VI} C_6^{VII})$	
I—V—VI	$A_3^I B^V C_4^{VI}$	$B_2^V C_5^{VI}$	III—IV—VI	$A_2^{III} B^{IV} C^{VI}$	$A_2^{III} C_3^{VI}$
I—V—VII	$A^I C^{VII}$	$B^V C_5^{VII}$	III—IV—VII	$A_3^{III} B_2^{IV} C^{VII}$	$A^{III} C_3^{VII}$
I—VI—VII	$A^I C^{VII}$	$B^{VI} C_6^{VII}$	III—V—VI	$A^{III} B^V$	$A_2^{III} C_3^{VI}$
II—III—VI	$A^{II} C^{VI}$	$B_2^{III} C_3^{VI}$		$(A_2^{III} C_3^{VI})$	$(B_2^V C_5^{VI})$
II—III—VII	$A_4^{II} B^{III} C_3^{VII}$	$A^{II} C_2^{VII}$	III—V—VII	$A^{III} B^V$	$A^{III} C_3^{VII}$
II—IV—VI	$A^{II} C^{VI}$	$B^{IV} C_2^{VI}$	III—VI—VII	$A_2^{III} B_3^{VI}$	$A^{III} C_3^{VII}$
II—IV—VII	$A_3^{II} B^{IV} C_2^{VII}$	$A^{II} C_2^{VII}$			

and

$$8 - C < B. \tag{22}$$

The expression (4) transforms into

$$x = \frac{8 - C}{2B}, \tag{23}$$

and Eq. (3) still applies. If we assume the condition $B \gtrless C$, it follows from Eqs. (21), (22), and (23) that there are 28 types of binary-defect compound satisfying Goryunova's rules (Table 10). The majority of these 28 types, given in Table 10, has representative compounds which may not exhibit defect structure but frequently have the tetrahedral pattern in the space lattice [19].

In addition to ternary compounds having diamond-like structure, in which the numbers of cations and anions in the compound are equal and the average number of valence electrons per atom is four, there are many other theoretically possible combinations of compounds resulting in the formation of ternary compounds which are crystallochemically similar to zinc blende but have an excess or a deficiency of some component.

In the case of defect compounds, it is usual to consider cases in which $\frac{1}{4}$, $\frac{1}{3}$, $\frac{1}{2}$, $\frac{2}{5}$, and $\frac{2}{3}$ of the metal sites in the structure of zinc blende exhibit valence properties. The average number of electrons per atom is not equal to four but varies from 4.57 to 6.00. According to Goryunova [12], the upper limit (which is probably too high) for the existence of tetrahedral structures is six electrons/atom. As demonstrated in [5], structures with 50% of vacancies are rare and those with 75% of vacancies represent discrete molecular clusters.

In spite of the fact that, in theory, an enormous number of defect ternary compounds may exist, the majority has not yet been found and even those which are already known are frequently found to be variable composition phases which are solid solutions of simpler compounds.

If we approach ternary defect compounds in the same way as binary defect compounds, we must use Goryunova's formulas for four-component phases since a vacancy can be regarded as an atom of zero valence. The simultaneous solution of normal valence equations and of the four-electron condition in the presence of three unknowns does not yield a definite result.

We can show that all the ternary defect phases deduced by Goryunova decompose into normal (four electrons/atom) binary or ternary phases or into the binary defect phases listed in Table 10 (in a few cases, they decompose into two binary defect phases). Table 11 lists groups of elements which form ternary defect phases (for $A < 4$) and components into which these phases decompose.

It is natural to assume that, in those cases when the components of a ternary defect phase have similar structures, solid solutions can be formed between them and the ternary phase then becomes either a solution of defect composition or it acquires a superstructure in the ordered state. Hence, we may expect that the $A^I B_5^{III} C_8^{VI}$ phase predicted by Goryunova should exist because they decompose into well-known $A^I B^{III} C_2^{VI}$ and $2B_2^{III} C_3^{VI}$ compounds which have similar crystal structures [20]. The formation of $A_6^{III} B^{IV} C_7^{VI}$ phases is also very likely because it represents a combination of $A_2^{III} B^{IV} C^{VI}$ (cf. Chapter V) and $2A_2^{III} C_3^{VI}$ compounds.

Crystal Structures of Ternary Semiconducting Compounds

A characteristic feature of diamond-like compounds is the tetrahedral coordination, whose appearance is usually related to the nature of the hybridization of the valence electron orbitals.

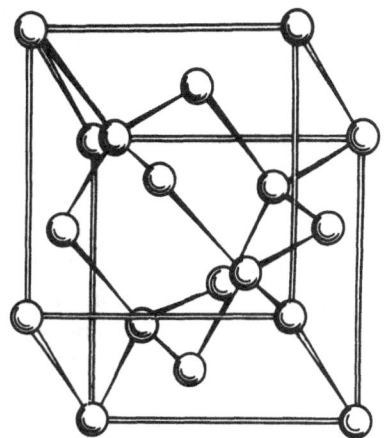

Fig. 8. Diamond structure.

The crystal structure of elements of the IVB subgroup (the diamond subgroup) can be represented by spheres packed in a face-centered lattice, in which half the tetrahedral vacancies are occupied by atoms of the same element (Fig. 8) or it can be represented by two superimposed face-centered cubic lattices formed by atoms of one kind. The space group of the diamond structure is $Fd3m(O_h^7)$.

When we consider binary compounds instead of elements, we find that the tetrahedral coordination can appear if the sp^3 hybridization is retained but, in this case, two space lattices may exist: they are the sphalerite or wurtzite lattices (Fig. 9). Sphalerite and wurtzite are the two crystallographic modifications of zinc sulfide; the former has a cubic symmetry with the space group $F\bar{4}3m(T_d^2)$, and the latter has a hexagonal symmetry with the space group $C6mc(C_{6v}^4)$. The diamond and sphalerite structures are described by the same Bravais lattice, which is a face-centered cube.

We must mention that the causes of the appearance of the tetrahedral coordination are not yet clear. For example, it is known that, in addition to $A^{III}B^V$, $A^{II}B^{VI}$, and $A^I B^{VII}$ compounds, the sphalerite structure is exhibited also by manganese sulfide and selenide, by titanium borides $TiB_{0.9-1.5}$, as well as by a large group of interstitial phases: β-TiH, δ-ZrH, NbH, PdH, TiH_2, CrH_2, and β-Zr_4H, although these substances do not obey Goryunova's rules.

In the case of ternary diamond-like semiconductors we have atoms of two elements in the cation (or anion) part of the sphalerite (or wurtzite) lattice. Atoms can be distributed in the cation or anion sublattice in two ways: ordered and disordered (random). In the latter case, a compound has a lattice similar to that of sphalerite (or wurtzite), while in the former case, we observe a tetragonal distortion of the cubic lattice, resulting from the ordered distribution of atoms of two different sizes in the cation (or anion) sublattice. In this case, the sphalerite structure is replaced by the tetragonal lattice of chalcopyrite (or antichalcopyrite); the formation of a nonrectangular lattice (for example, that of the enargite or famatinite type) is possible. The chalcopyrite structure is shown in Fig. 10.

The appearance of the chalcopyrite structure due to the ordering of atoms of two kinds in one of the sphalerite sublattices has been confirmed experimentally: many of the ternary

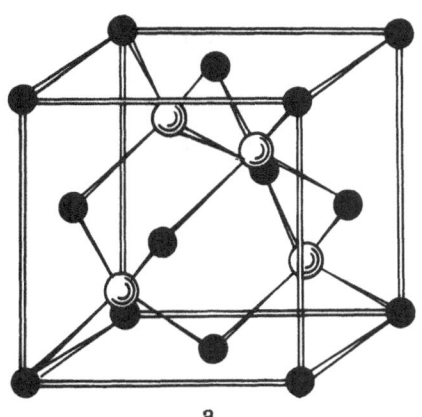

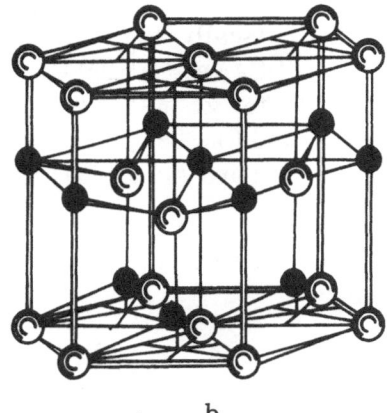

a   b

Fig. 9. Structures of sphalerite (a) and wurtzite (b).

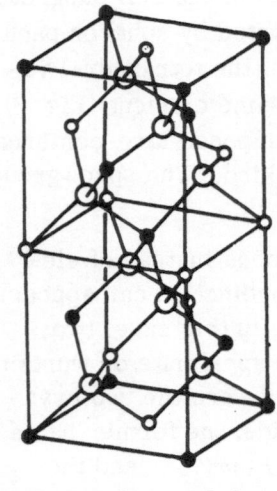

O Sn ● Cd O As

Fig. 10. Structure of
chalcopyrite.

compounds which have the chalcopyrite structure at a sufficiently low temperature, exhibit endothermic effects during heating, which indicate a phase transition in the solid state [21]. X-ray diffraction investigations of the high-temperature phases of some ternary compounds have demonstrated that they have the sphalerite structure.*

The cause of the existence of a metastable high-temperature phase follows in a natural manner from thermodynamic considerations: when the temperature is increased, the free energy becomes larger and the probability of a random distribution of atoms of different kinds' in the cation (or anion) sublattice of chalcopyrite (or antichalcopyrite) increases without disturbances of the tetrahedral coordination because the interaction of atoms in one sublattice is much weaker than their interaction with atoms in the other sublattice (due to the small difference between their electronegativities). The temperatures of the chalcopyrite-sphalerite phase transitions will be given later in a chapter which describes these compounds (Chapter IV).

The problem of the ordering of atoms in multicomponent semiconductors has not yet been studied sufficiently thoroughly. The microscopic ordering mechanism in ternary semiconductors has been considered first by Folberth and Pfister [23], who have attributed such ordering to the difference between the polarizabilities of bonds. Palatnik et al. [24] have pointed out that the conclusions of Folberth and Pfister agree with the experimental results only in the case of $A^{II}B^{IV}C_2^{V}$ compounds but not in the case of other compounds. As a measure of ordering, Palatnikov et al. [24] suggest the quantity

$$\delta = a - c/2$$

(where a and c are the lattice parameters) and, using a large group of $A^{I}B^{III}C_2^{VI}$ and $A^{II}B^{IV}C_2^{V}$ compounds, they show that δ is proportional to the product of the difference between the electronegativities of the components and the difference between their ionic radii. However, it is more convenient to use a dimensionless parameter, for example, δ/a, as a measure of ordering because it is not related, for a given tetrahedral distortion, to the unit cell parameters.

We shall consider in more detail the characteristic features of the chalcopyrite structure. It can be regarded as the sphalerite structure doubled along the c direction [35], with the metallic part consisting of atoms of two kinds. Such a modification of the sphalerite lattice distorts it somewhat. Usually, the distortion is weak and tetragonal: in known ternary semiconductors of the $A^{I}B^{III}C_2^{VI}$ and $A^{II}B^{IV}C_2^{V}$ type the ratio of the c/a axes varies from 1.80 for $AgAlSe_2$ [31] to 2.00 for $ZnSnAs_2$ [23]. In general, other distortions are also possible. For example, Cu_3AsS_4 and Cu_3SbS_4 [32] have orthorhombic lattices with the tetrahedral coordination while $BeSiN_2$ [33] and $AgInS_2$ [31] have the wurtzite-type structures.

The chalcopyrite structure can be considered in detail by analyzing the unit cell of $CdSnAs_2$ [29, 30]. It contains four formula units. Atoms in this lattice are located at the following positions:

As: $(x, a/4, c/8)$, $(-x, 3a/4, c/8)$,

$(3a/4, x, 7c/8)$, $(a/4, -x, 7c/8)$, $(x < a/4)$;

---

*Yu. V. Rud', Dissertation for Candidate's Degree [in Russian], Physicotechnical Institute, Academy of Sciences of the USSR, Leningrad (1965).

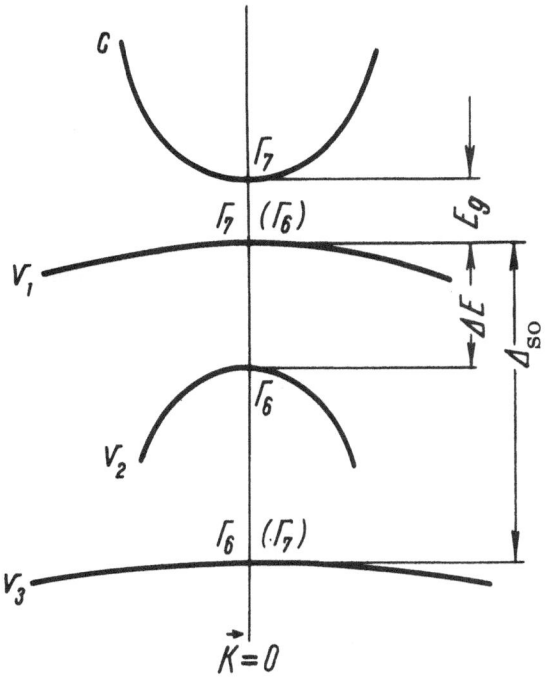

Fig. 11. Band structure of a chalcopyrite-type lattice.

Cd: (0, 0, 0) and (0, $a/2$, $c/4$);

Sn: (0, 0, $c/2$) and (0, $a/2$, $3c/4$).

These points form a tetragonal body-centered translation lattice with edges a, a, c. The Bravais lattice is represented by the matrix:

$$A = \begin{pmatrix} -a/2 & a/2 & a/2 \\ a/2 & -a/2 & a/2 \\ c/2 & c/2 & -c/2 \end{pmatrix},$$

which, multiplied by a unit integer vector $\mathbf{i}$, generates the whole crystal. The reciprocal lattice consists of points whose coordinates are $B\mathbf{j}$, where $\mathbf{j}$ is a unit integer vector, and

$$B = 2\pi \begin{pmatrix} 0 & 1/a & 1/c \\ 1/a & 0 & 1/c \\ 1/a & 1/a & 0 \end{pmatrix}$$

represents the tetragonal face-centered lattice with edges $4\pi/a$, $4\pi/a$, and $4\pi/c$.

This polyhedron has the symmetry of a prism with a square base (D_{4h}): it has one four-fold axis (001) and four twofold axes (100), (010), (110), and ($\bar{1}$10); it exhibits inversion as well as inversion-rotation combinations. Of 16 possible operations, only half (rotation and mirror reflections) apply to the point group of the chalcopyrite structure D_{2d}. Its space group $I\bar{4}2d$ (D_{2d}^{12}) is nonsymmorphic and half of its elements are related to the nonprimitive translations $\tau = (0, a/2, c/4)$.

TABLE 12. True Covalent Tetrahedral Radii, nm(Å)

period	Group						
	IB	IIB	IIIB	IVB	VB	VIB	VIIB
2					N 0.07 (0.70)		
3		Mg 0.14 (1.40)	Al 0.1317 (1.317)	Si 0.117 (1.17)	P 0.11 (1.10)	S 0.104 (1.04)	
4	Cu 0.136 (1.36)	Zn 0.139 (1.39)	Ga 0.131₅ (1.31₅)	Ge 0.122 (1.22)	As 0.117 (1.17)	Se 0.114₅ (1.14₅)	Br 0.12 (1.20)
5	Ag 0.155 (1.55)	Cd 0.156₅ (1.56₅)	In 0.148₅ (1.48₅)	Sn 0.14 (1.40)	Sb 0.135 (1.35)	Te 0.128 (1.28)	I 0.113₄ (1.13₄)
6		Hg 0.154₅ (1.54₅)	Tl 0.154₂ (1.54₂)				

The band structure corresponding to the chalcopyrite lattice was investigated theoretically by Gashimzade [34] as well as by Sandrock and Treusch [29]. The band structure at the point $\mathbf{k} = 0$, where the maximum of the valence band is located, is shown in Fig. 11 [36]. The upper valence band v_1 (the heavy-hole band) may or may not have the same symmetry as the conduction band c. According to the characteristic features of the symmetry of the v_1 band, the spin-orbit split valence band v_3 may or may not have the same symmetry as the conduction band c. The v_2 band is the light-hole band, which is split because of a reduction in the symmetry on transition from the sphalerite to the chalcopyrite structure. The gap Δ_{so} between the valence bands v_1 and v_3 is due to the spin-orbit splitting.

Covalent radii, calculated by Pauling and Huggins [26], can be used to calculate the lattice periods in compounds with a predominantly covalent type of binding. However, in the case of compounds with the chalcopyrite structure, it is found that the lattice parameters calculated in this way do not agree with the experimental data.

The task of refining the covalent radii was undertaken by Palatnik et al. [25]. They excluded the ionic components of the binding forces and obtained the values of what they call the true covalent tetrahedral radii for some elements of the B subgroups in groups I-VII of the periodic system (which are listed in Table 12). Palatnik et al. [25] calculated the average interatomic distance by means of the following formula, which is an empirical formula for binary compounds [27] generalized to the case of multicomponent compounds:

$$\bar{d} = 2\Sigma_i r_i \eta_i - 0.09 \Sigma_{ik} \left| (\Delta x)_{ik} \right| \varepsilon_{ik},$$

where r_i is the radius of the i-th component, taken from Table 12; η_i is the atomic concentration of the i-th component in the compound; $(\Delta x)_{ik}$ is the difference between the electronegativities [28] of the components i and k forming a bond; and ε_{ik} is the ratio of the number of bonds between the components i and k to the total number of bonds.

The values of the lattice parameters calculated in [25] [the average parameter for the chalcopyrite structure is assumed to be $\bar{a} = \frac{1}{3}(2a + c/2)$] for a large number of binary and two-cation ternary compounds with the sphalerite and chalcopyrite structures are in good agreement with the experimental data.

CHAPTER III

METHODS FOR THE PREPARATION OF
TERNARY COMPOUNDS

Synthesis and Crystallization of Ternary Compounds

The purity and perfection requirements which semiconducting materials must satisfy are so stringent that the methods of synthesis of ternary compounds, based on the direct fusion of elements, are frequently unsuitable.

Most of the phase diagrams of ternary systems, in which ternary semiconducting compounds with the tetrahedral coordination are formed, have not yet been investigated. It has been pointed out in [37] that the nature of the phase diagram of a ternary system must be taken into account in the selection of the synthesis method because the available published data suggest that ternary compounds frequently melt incongruently.

The conclusion reached in [37] is far too sweeping but it is in agreement with the available published data. It has been reported in [38] that $ZnSnAs_2$, formed in the $Zn-Sn-As$ system, melts in accordance with a peritectic reaction. The $T-X$ sections of the phase diagram of this system are given in Fig. 12.

Investigations of selenides and tellurides of the $A^{II}B_2^{III}C_4^{VI}$ have established that the majority of these compounds also melts peritectically [39] (cf. Chapter V). Moreover, peritectic melting has been observed for several other ternary compounds which do not have the tetrahedral coordination.

The development of the synthesis methods for ternary compounds containing volatile components (such compounds form the majority of those considered in the present monograph) is very greatly complicated by the fact that, in this case, we have to know not only the $T-X$ sections of the phase diagrams but also the full $P-T-X$ diagrams because the vapor pressure above the melt is considerable. In other words, it is necessary to investigate the dependence of the phase composition on the pressure, temperature, and chemical composition.

It has been shown in [40, 41, 56] that in order to prevent the dissociation of a compound it is necessary to establish a vapor pressure of the volatile element equal to the dissociation vapor pressure of the compound at its melting point. When the vapor pressure in the system exceeds the dissociation pressure the compound melts congruently but the melt obtained contains an excess of the volatile component and is not stoichiometric. When the vapor pressure of the volatile component is too low, the compound melts incongruently, i.e., it decomposes into a liquid enriched with the nonvolatile component and a vapor of the volatile component.

It follows that it is essential to know the $P-T-X$ phase diagram in the direct synthesis of ternary compounds.

Any deviation of the composition of a ternary compound from the stoichiometric ratio may affect very considerably its semiconducting properties. By way of example, we can quote

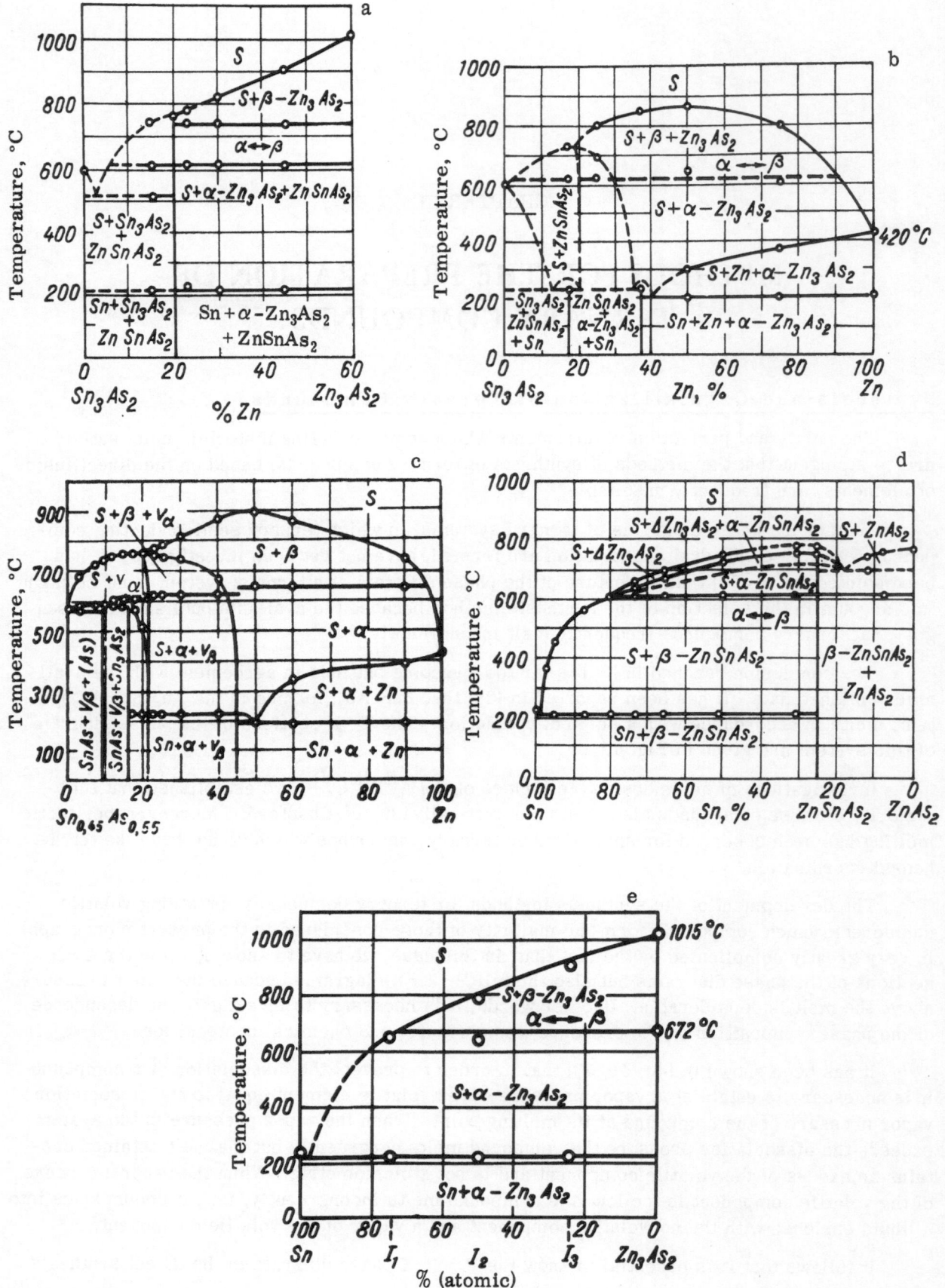

Fig. 12. T—X sections of the Zn—Sn—As system.

a change in the resistivity of $CdIn_2S_4$ (cf. Chapter V) by several orders of magnitude when the synthesis of this compound is carried out under sulfur vapor pressure. Therefore, in investigations of semiconducting properties, we can obtain reliable data only when the samples are homogeneous and stoichiometric. The considerable discrepancies between the published values of the properties of ternary compounds are most likely to be due to a lack of homogeneity and stoichiometry.

Another factor which complicates direct synthesis of some ternary compounds is their polymorphism. It has been established [42]* that, for example, the compounds $CdSnAs_2$ and $ZnGeAs_2$, which have the chalcopyrite structure at low temperatures, undergo a polymorphic transition to the zinc-blende structure near the melting point. In this case, thermograms show an additional thermal effect and the zinc-blende lattice parameter assumes an intermediate value between the parameters a and $c/2$ of the chalcopyrite structure. We may assume that the large number of cracks in ingots of, for example, $CdSnAs_2$, prepared by direct synthesis from elements (such cracks make these ingots unsuitable for electrical measurements) is not only due to a difference between the thermal coefficients along the a and c directions of the chalcopyrite structure but also due to a polymorphic transition.

We must mention also another limitation of the use of the direct-synthesis method in the preparation of ternary compounds in evacuated quartz ampoules. This limitation refers to the synthesis of ternary compounds having a high melting point. By way of example, we can quote here such compounds as $BeSiN_2$, $ZnSiP_2$, and $ZnGa_2S_4$, whose melts cannot be prepared in quartz ampoules.

We must thus conclude that the majority of ternary compounds prepared by direct synthesis from the component elements, using the available techniques and the very limited information on the phase diagrams, cannot be perfect in respect of their composition and homogeneity.

This does not mean that direct synthesis of ternary compounds from their elements should not be tried. It should be used to prepare ingots or sintered aggregates which are then recrystallized by various methods which will be described in subsequent sections. Moreover, studies of ingots prepared by direct melting can yield some preliminary information on the properties of ternary compounds.

One of the considerable difficulties in the synthesis of compounds is to prevent the contamination of the melt.

Each ampoule should be etched in hydrofluoric acid in order to clean the internal surface and remove small projections which might act as spontaneous crystallization centers; this should be followed by washing in distilled water in order to remove the traces of the acid (at this stage, it is desirable periodically to check the electrical conductivity of the water after each washing). The next stage is to remove traces of organic compounds from the surface of the quartz by means of acetone or rectified ethyl alcohol. Next, the ampoule should be washed again in distilled water and dried in a sterile vacuum drying chamber.

When quartz is used as the crucible material, the melt may be contaminated by impurities present in quartz, particularly with gases adsorbed on its walls. It has been reported in [44] that the main component of the mixture of gases evolved by molten quartz at 600°C is water but at 1150°C it is replaced by hydrogen. Our own experiments confirmed indirectly this conclusion: when ternary compounds containing sulfur, selenium, or tellurium were prepared, a strong smell of H_2S, H_2Se, or H_2Te was experienced when the ampoules were opened after cooling.

---

*See also É. O. Osmanov, Dissertation for Candidate's Degree [in Russian], Institute of Silicate Chemistry, Leningrad (1965).

TABLE 13. Temperature Dependences of Vapor Pressures and Melting Points of Some Elements

Element	Temperatures (°C) corresponding to vapor pressures in N/m² (mm Hg)					M. P., °C
	1330 (10)	5320 (40)	13300 (100)	53200 (400)	101080 (760)	
Ag	1575	1743	1865	2090	2212	960.5
Al	1487	1635	1749	1947	2056	660
As	437	483	518	579	610	814
Bi	1136	1217	1271	1370	1420	271
Cd	484	553	611	711	765	320.9
Cu	1879	2067	2207	2465	2595	1083
Ga	1541	1680	1784	1974	2071	30
Pb	1162	1309	1421	1630	1744	327.5
P (red)	287	323	349	391	417	590
Sb	1033	1141	1223	1364	1440	630.5
S	243.8	288.3	327.2	399.6	444.6	112.8
Se	442	506	554	637	680	217
Sn	1703	1855	1963	2169	2270	231.9
Te	650	753	833	997	1037	452
Tl	983	1103	1196	1364	1457	303.5
Sn	593	673	736	844	907	419.4

The problem of the influence of the container (crucible) material on the semiconducting properties of ternary compounds has not yet been investigated sufficiently thoroughly.

The accuracy of weighing, the purity of air in the laboratory, etc., must also satisfy stringent requirements.

Ternary compounds whose melting points do not exceed 1200°C can be prepared in quartz ampoules [46, 47, 49]* using vibrational stirring [50]. Usually the initial charge is heated rapidly to the melting point of the most easily fusible element and kept at this temperature for some time; the temperature is then gradually raised to a level 50-100 deg above the melting point of the ternary compound and is kept at this level for 2-48 h. The cooling rate may affect considerably the properties of the ingots obtained: when this rate is reduced, the quality of the samples obtained increases.

In view of the fact that most of the P−T−X diagrams of the ternary systems have not been investigated, some workers [46, 173] have simply added to the ampoule an excess of the volatile component. The synthesis of some $A_3^I B^V C_4^{VI}$ compounds has been carried out [173] under a nitrogen pressure of 0.66 atm with an excess of 0.5 g of arsenic, selenium, or sulfur added to a charge of 50 g. However, an investigation of $CuInSe_2$ [89], which has a considerable vapor pressure near the melting point, showed that the losses of the volatile component cannot be compensated by counterpressure because the vapor above the melt has a complex composition.

If the melting point of a ternary compound is high and its components have high vapor pressures (this applies to phosphorus, arsenic, sulfur, etc.), synthesis at one temperature may result in an explosion. In this case, it is preferable to use some variant of the two-temperature method [41, 51, 53].

The vapor pressure of a volatile component should be controlled. Table 13 lists the vapor pressures and melting points of elements which occur in ternary compounds [54]. The selection of a particular synthesis technique should be based on the total vapor pressure of all the elements in the compound.

---

*See also Dissertations for Candidate's Degree submitted by S. Mamaev (Gertsen Leningrad Pedagogical Institute, 1962) and V. M. Koshkin (Khar'kov State University, 1964).

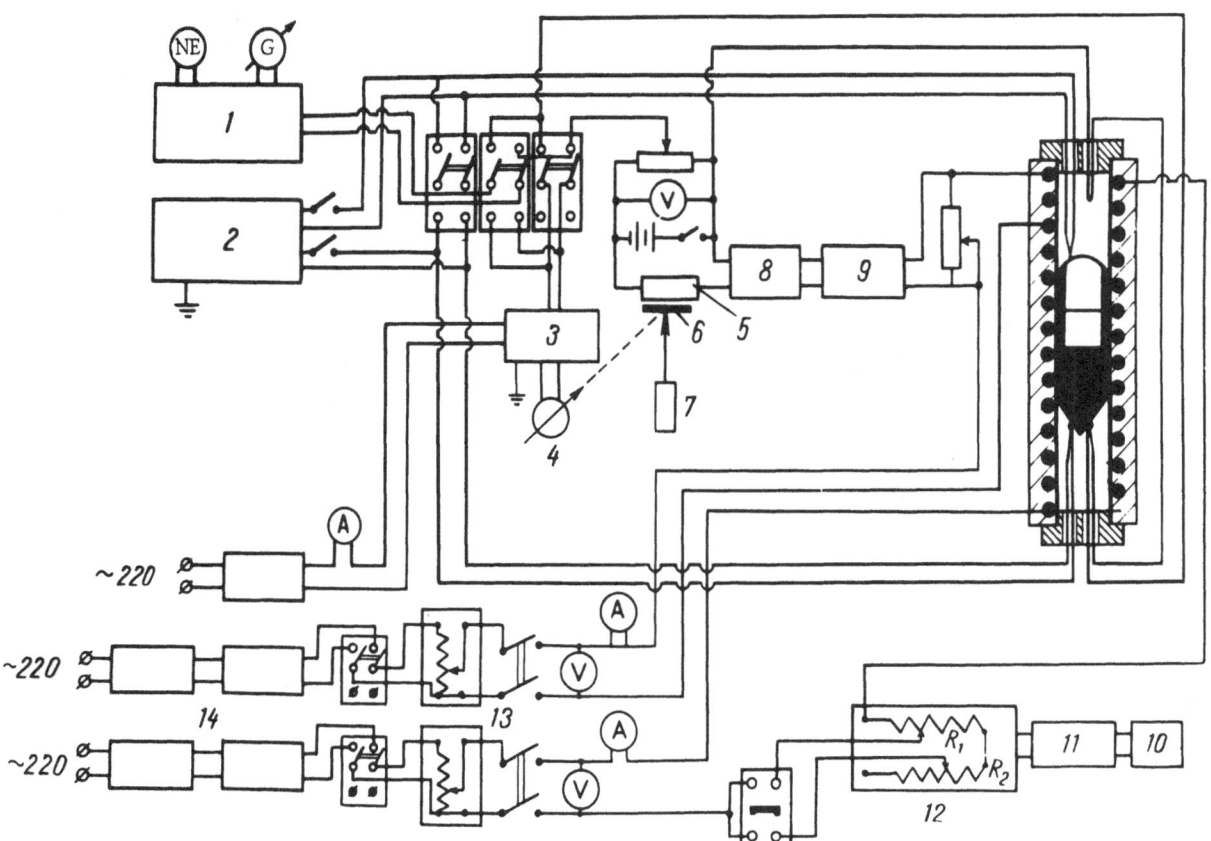

Fig. 13. Basic layout of the apparatus with a constant temperature gradient: 1) PPTV-R300 potentiometer; 2) ÉPP-09 potentiometer; 3) F115/V-1 amplifier; 4) GMP null galvanometer; 5) photoresistor; 6) screen; 7) light source; 8) RP-7 relays; 9) contactor; 10) RD-09 motor; 11) reduction gear; 12) rheostats; 13) RNO-250-2 autotransformers; 14) voltage stabilizers.

Some ternary compounds have been prepared using the methods for growing crystals from the melt, which have been developed for elemental and binary semiconductors. These methods include the Bridgman method, directional cooling, and zone recrystallization.

The application of the Bridgman method to the growth of $CdSnAs_2$ crystals did not give satisfactory results: the ingots had many cracks [21]. Similar results have been reported in [43] for crystals of other $A^{II}B^{IV}C_2^{V}$ compounds.

The pulling of a melt-filled ampoule from a furnace at a rate of 3 cm/h has been employed in the preparation of coarse-grained crystals of ABX_2 compounds [94]. However, even in this case it has not been possible to grow large crystals [94], due to the anisotropy of the linear expansion coefficient. The same explanation is given in [89] for the lack of success in the growth of large crack-free crystals of ABX_2 compounds by pulling an ampoule from a furnace, by directional cooling, or by zone recrystallization.

Good results have been reported in [43], which describes a method for the preparation of single crystals of the compounds $ZnGeAs_2$, $ZnSnAs_2$, and $ZnSiAs_2$. A special apparatus has been developed by means of which directional cooling can be applied at a constant temperature gradient [55]. The furnace has, in addition to the main winding, an auxiliary winding which is used to establish temperature gradients from 1 to 15 deg C/cm by varying the current in this winding. The basic layout of the apparatus is shown in Fig. 13.

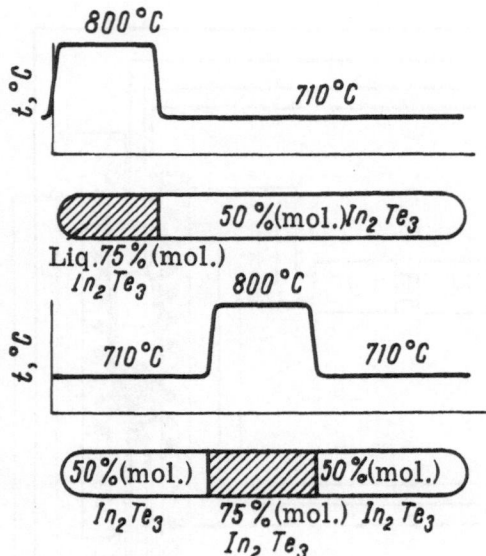

Fig. 14. Temperature distribution in a furance during zone melting [277].

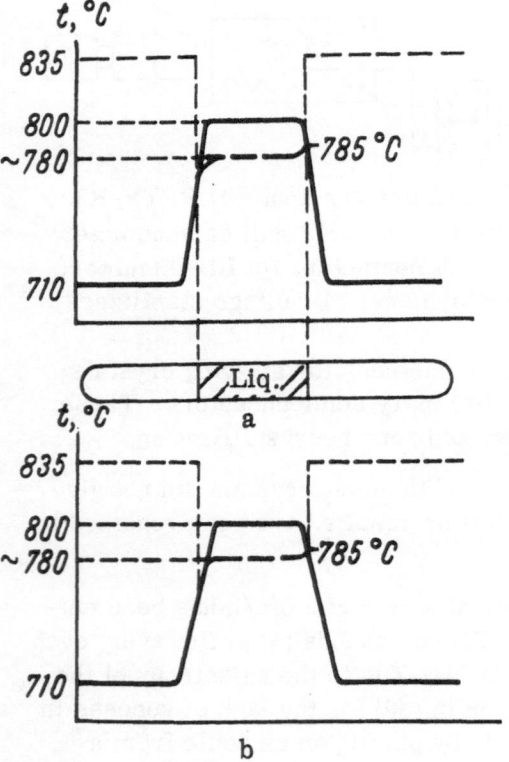

Fig. 15. Stability criteria of a solidifying surface of CdIn$_2$Te$_4$: a) stable solidification; b) unstable solidification.

The method of zone recrystallization has been used to prepare single crystals of ternary compounds. It is reported in [173] that compounds of the $A_3^I B^V C_4^{VI}$ type, having low melting points and low values of the vapor density at temperatures exceeding the melting point by 100 deg, can be easily purified by zone recrystallization and obtained in the form of single crystals. However, the present authors are not aware of successful experiments resulting in the growth of single crystals of such compounds by the zone recrystallization method. On the contrary, the few published reports indicate that the zone recrystallization of ternary compounds should be used with caution when the phase diagram is not known. Thus, for example, zone recrystallization as well as directional cooling of CuInSe$_2$ has been found to alter the lattice parameters along the ingot, indicating a change in the composition of the ingot [89]. Positive results, reported in several papers [108, 151, 155, 157], cannot be explained yet on theoretical grounds.

Only one example is known [277] of zone recrystallization of a ternary compound, based on the knowledge of the phase diagram of solid solutions, which includes the ternary compound. This example is the growth of CdIn$_2$Te$_4$, based on a system of CdTe$-$In$_2$Te$_3$ solid solutions, which melts in accordance with a peritectic reaction. The phase diagram of this system is given in Fig. 41. Two variants of the zone recrystallization of CdIn$_2$Te$_4$ are reported in [277]. In the first variant, the authors used a molten zone containing more than 63 mol.% In$_2$Te$_3$. The temperature distribution in the furnace is shown in Fig. 14. The temperature of the liquid zone was about 800°C, i.e., it was higher than the peritectic temperature of 785°C. The temperature of the other parts of the sample was maintained at 710°C, which was higher than the temperature of the next transformation at 702°C and which, therefore, prevented any precipitation of the γ-phase.

In the second variant, the authors used an ingot containing 53 mol.% In$_2$Te$_3$. The first pass of the molten zone precipitated the α-phase, saturated with In$_2$Te$_3$, and this continued until the concentration of In$_2$Te$_3$ in the molten zone reached 63 mol.%. Next, the β-phase was formed and the concentration of In$_2$Te$_3$ in the molten zone continued to increase. During the second pass (in the opposite direction), the concentration of In_2Te_3 decreased because the α-phase was dissolved. During subsequent passes, the amount of the α-phase decreased until it disappeared completely. In this way, several passes of the zone in both directions were sufficient to prepare homogeneous samples containing 50 mol.% In_2Te_3.

It was found [277] that the precipitation of In_2Te_3 at the crystallization surface reduced the liquidus temperature and the dissolution of the substance at the other end of the zone increased this temperature. When the temperature gradient in the melt was higher than the equilibrium gradient of the liquidus temperature, based on the concentration gradient, the growth of the solidifying surface became stable (Fig. 15a). Otherwise, the unstable growth of dendrites was observed and inclusions were captured by the ingot (Fig. 15b). The methods for the growth of crystals of ternary compounds from the melt are thus not yet fully developed.

Preparation of Ternary Compounds from Molten Solutions

Semiconducting crystals can also be grown from molten solutions. The solvent may be the melt of one of the components of a compound or some other substance which does not react chemically with the compound being grown.

Growing of crystals from molten solutions is a very promising method in the case of refractory compounds or compounds having a high vapor pressure at the melting point or those which melt incongruently, because growth from a molten solution can be carried out below the melting point of the compound being prepared.

The main requirements which the solvent must satisfy were formulated in [56, 57].*

The solvent should:

1)   dissolve the substance being grown in a sufficient amount;
2)   reduce considerably the crystallization temperature at fairly high concentrations;
3)   have low value of the saturated vapor pressure in the working range of temperatures;
4)   remain liquid in a wide range of temperatures;
5)   have no appreciable effect on the properties of the compound being grown and have a slight solid-state solubility in the crystals being grown;
6)   be easily separable from the grown crystals;
7)   be chemically inert with respect to the ampoule or the crucible.

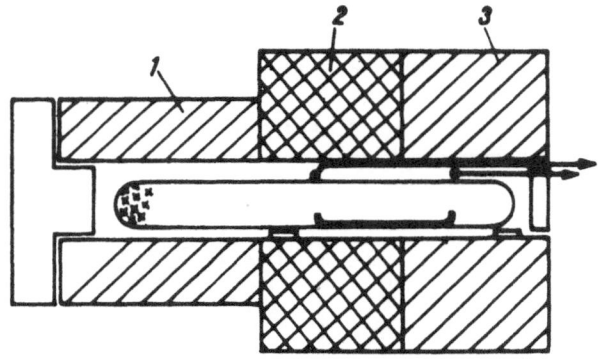

Fig. 16. Apparatus for the growth of crystals from a molten solution: 1) furnace used to heat B^V; 2) low-temperature furnace; 3) high-temperature furnace.

---

*See also É. O. Osmanov, Dissertation for Candidate's Degree [in Russian], Institute of Silicate Chemistry, Leningrad (1965).

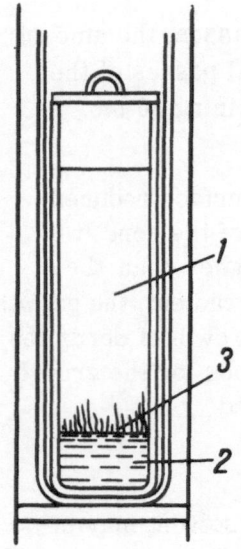

Fig. 17. Apparatus used to grow ZnS crystals: 1) molten salt; 2) zinc sulfide; 3) growing crystals.

The methods based on recrystallization from a solution in the melt of one of the components of a compound are usually employed for binary compounds with a known phase diagram. In such cases, crystals of a compound can be obtained at a relatively low temperature, as indicated by the liquidus curve near the pure metallic component.

This method has been used [58] to prepare $A^{III}B^V$ crystals (A^{III} = Ga, In, Al; B^V = P, As, Sb); crystals are prepared by cooling the melt at a rate such that the crystallization velocity remains constant. The experimental techniques have since been improved. Figure 16 shows apparatus for growing crystals from a molten solution [59]. A temperature gradient is established along a boat (containing molten A^{III}) by the use of two furnaces; the vapor pressure of B^V is controlled by a third furnace. This method has advantages over the method described in [58], because large ingots can be obtained.

Osmanov (cf. his dissertation) was the first to grow ternary compounds from molten solutions. Zinc and cadmium were used as the solvents for $ZnSiP_2$ and $CdSiP_2$. However, the high vapor pressures of these elements at elevated temperatures presented technical difficulties, while at low temperatures the solubility was insufficient and the crystals were obtained in the form of dendrites. Therefore, Osmanov tried other solvents. He found tin to be most effective. He introduced 10-20 mol.% of the ternary compound into the melt and, after rapid heating to 1250°C, he cooled the melt at a rate of 20 deg/h.

$ZnSiP_2$ and $CdSiP_2$ crystals were in the form of elongated prisms. When the rate of cooling was altered, the habit of the crystals changed. Thus, for example, when the rate of cooling was 50 deg/h, the crystals obtained were thin and long, and when the cooling rate was 5 deg/h, the crystals were isometric.

$ZnSiP_2$ and $CdSiP_2$ compounds could be recrystallized also from molten lead, indium, bismuth, and antimony.

It has been reported in [62] that twinned $ZnSnAs_2$ plates, with very well-formed faces, were obtained from molten tin. The melt, containing 1.4% Zn, was cooled from 800°C at a rate of 1.3 deg/min.

However, not all the ternary arsenides can be prepared from a solution in one of its components. Thus, the present authors attempted to prepare $CdSnAs_2$ crystals from tin but ob-

TABLE 14. Solvents and Results of Recrystallization

Solvent	t_{mp}, °C	t_{bp}, °C	Size of crystals, mm	t_{growth}, °C
NaCl	803	1443	Needles, 0.5 × 10	1070
KCl	778	1414	Plates, 5 × 0.2	1200
NaBr	766	1393	Needles, 0.2 × 5	1050
NaI	661	1300	Needles, 0.05 × 2	1050
KI	686	1324	Needles, 0.05 × 2	1050
$CaCl_2$	772	1600	Needles, 0.1 × 5	1050
LiCl	614	1357	No crystals	1050
KBr	748	1378	No crystals	1050
$ZnCl_2$	262	730	No crystals	550
$CdCl_2$	568	960	No crystals	750
Na_2CO_3	852	—	No crystals	1050
Na_2SO_4	884	1702	No crystals	1050

tained instead a binary compound of tin and arsenic. The following reaction takes place when $A^{II}B_2^{III}C_4^{VI}$ compounds are crystallized from molten B^{III}:

$$A^{II}B_2^{III}C_4^{VI} + B_{\text{melt}}^{III} \longrightarrow A^{II}C^{VI} + B^{III}C^{VI}.$$

In other words, when, for example, $ZnGa_2S_4$ is introduced into molten gallium, ZnS and GaS crystals are obtained. The recrystallization of $A^{II}B_2^{III}C_4^{VI}$ compounds from molten elements of group IV produces crystals of binary compounds. Thus, for example, the recrystallization of ternary sulfides from tin produces SnS crystals [63].

One can also use compounds — in particular, various salts — as the solvents. Methods for the preparation of zinc sulfide crystals from molten salts are described in [57, 60]. The experiments were carried out in quartz ampoules with double walls. These ampoules were pumped out at 200°C in order to remove water vapor. The hot melt was kept at an appropriate temperature for 12-100 h in order to establish equilibrium conditions and then it was cooled at a rate of 2-10 deg/h. The apparatus used is shown schematically in Fig. 17. The most suitable solvent for the growth of zinc sulfide crystals is potassium chloride. Table 14, which gives the results reported in [57], lists various salts and the results of recrystallization of zinc sulfide.

It is evident from this table that a suitable solvent has to be found by trial and error. In view of the success in the recrystallization of binary sulfides from molten salts, one of the present authors and his colleagues tried to recrystallize ternary sulfides [63]. It was found that ternary chalcogenides of the $A^{II}B_2^{III}C_4^{VI}$ system could be recrystallized from molten salts containing the same cation. Thus, for example, $CdIn_2S_4$ could be recrystallized from molten cadmium chloride. However, the use of $ZnCl_2$ failed to achieve the recrystallization of zinc compounds.

An attempt to recrystallize $CdSnAs_2$ from molten cadmium chloride was also unsuccessful. Obviously, this system had eutectic point because the precipitate obtained from cadmium chloride contained practically no faceted crystals but consisted mainly of fine irregular aggregates, whose Debye diffraction patterns indicated that they represented the compound $CdSnAs_2$.

Methods for the preparation of crystals of compounds containing oxygen from melts of various salts and oxides have been published. For example, we can mention the preparation of barium titanate crystals which have the perovskite structure, from molten carbonates [64], or the growth of magnetic garnets of the $Me_3Fe_5O_{12}$ type from molten lead oxide [65]. However, these compounds do not have the tetrahedral coordination, and preparation of such crystals is outside the scope of the present book. We may assume that ternary defect compounds containing oxygen and having the spinel structure can be recrystallized from molten salts or oxides using methods such as those employed for barium titanate.

The preparation of crystals of ternary compounds from molten solutions is a very promising method but one which has not yet been investigated sufficiently. This is because of the time-consuming preliminary stage involving the determination of the phase diagrams of complex systems, which must be known before the growth of crystals from molten solutions can be attempted successfully.

Preparation of Ternary Compounds by Gas-Transport Reactions

The method of gas-transport reactions is being used increasingly widely in the preparation of single crystals of various compounds. This method is one of the variants for the growth of crystals from the gaseous (vapor) phase, but the gas-transport reactions have the advantage that they can take place at relatively low temperatures. This aspect is very important in the preparation of single crystals of refractory semiconductors. Moreover, gas-transport reactions

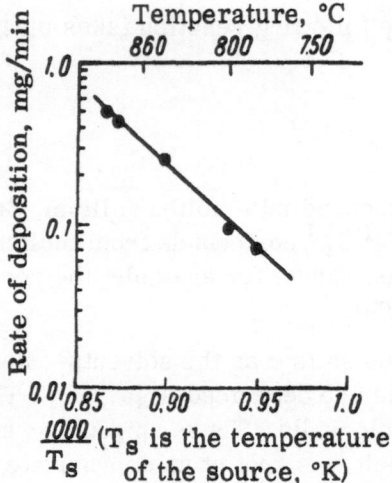

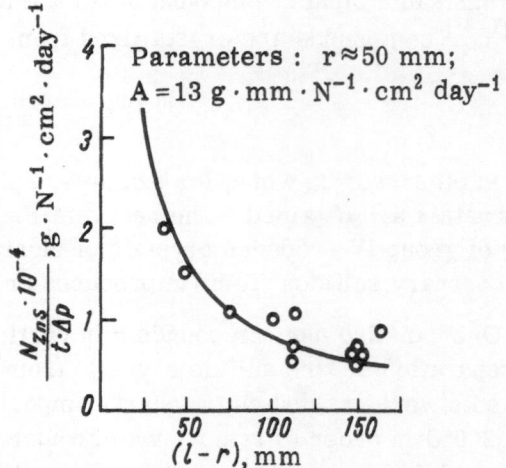

Fig. 18. Influence of the temperature of the source on the rate of deposition of gallium phosphide.

Fig. 19. Dependence of the rate of transport of zinc sulfide on the length of an ampoule.

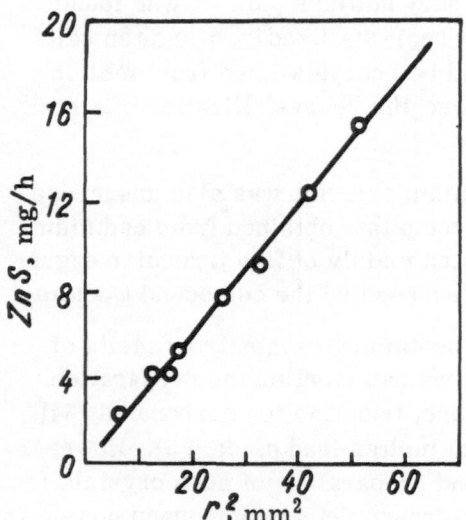

Fig. 20. Dependence of the rate of transport of zinc sulfide on the square of the radius of an ampoule.

are very promising in the growth of crystals of substances which melt incongruently, since the growth may take place at temperatures considerably lower than the melting point of the compound being prepared.

If a suitable gas is available, matter can also be transported when the saturated vapor pressure of the transported substance is slight. In this respect, the method of chemical transport reactions is more convenient than sublimation methods.

The transport takes place when a solid reacts reversibly with the transport gas, forming a volatile substance, which can then be decomposed to yield the original substance. The reaction must be reversible and a concentration gradient in the system is essential [66].

If a heterogeneous reaction between a solid and the transport gas is written in the general form $x_s = q_aA + q_bB + ... + q_rR + ...$, then, as demonstrated by Lever [67], the equilibrium constant of the reaction is related to other parameters by the expression

$$\Delta \ln K - \sigma = F\psi\Delta X,$$

where K is the equilibrium constant; σ is a quantity which represents supersaturation; $F = -F_x$ represents the rate of growth of crystals; $\psi = \frac{1}{2}\sum_s\sum_r\Phi_{rs}/ND_{rs}$; N is the total molar density; D_{rs} is the diffusion coefficient; $\psi_{rs} = (q_rp_s - q_sp_r)^2/p_rp_s$; p are partial pressures; and q are coefficients in the chemical reaction equation.

To ensure a sufficiently rapid transport of matter, the equilibrium should not be shifted too much toward either side.

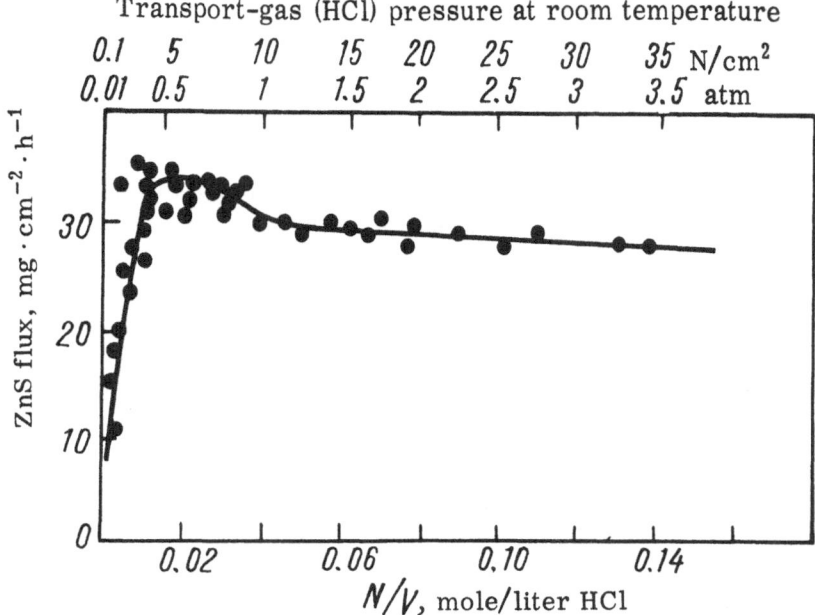

Fig. 21. Dependence of the rate of transport of zinc sulfide on
the concentration of the transport gas.

Using the expression $d \ln K / dT = \Delta H/RT^2$, we can determine the direction of the transport of matter from the sign of the enthalpy of the reaction. If the equilibrium constant increases with increasing temperature, the reaction is endothermic and the value of ΔH is positive so that the transport takes place from a hotter to a cooler zone; in the exothermic reaction case, the direction of the transport is opposite [66, 68].

The dependence of the rate of transport on various parameters has been determined theoretically in [67]. This dependence is given by the following formula:

$$F_s = \frac{\Delta H \Delta T}{R^2 \bar{T}^3 L} \left\{ \sum_i \frac{q_i}{p_i} \left(\sum_j \frac{q_j \bar{p}_i - q_i \bar{p}_j}{PD_{il}} \right) \right\}^{-1},$$

where F_s is the rate of transport through a unit cross-sectional area; $\Delta T = T_2 - T_1$; R is the universal gas constant; $\bar{T}$ is the average temperature; $\bar{p}_j$ and $\bar{p}_i$ are the average partial pressures of the components j and i; P is the total pressure; D_{ij} are the diffusion coefficients for the binary mixture of components i and j; and q are the coefficients in the chemical reaction equation.

It follows from this formula that the amount of transported matter depends on the average temperature, the temperature difference, the cross section and length of the ampoule, and the concentration of the transport gas. The influence of all these factors must be taken into account in the growth of crystals.

Many papers have been published on the gas-transport growth of binary compounds of $A^{III}B^V$ and $A^{II}B^{VI}$ type. These investigations can be used as the starting point in the growth of crystals of ternary compounds, because there have been few investigations of the gas transport growth of ternary crystals. We shall now consider individually the influence of various parameters on the transport of binary compounds. Figure 18 shows the influence of the source temperature on the rate of deposition of gallium phosphide at a constant pressure of the transport gas [69]. The influence of the geometrical factors can be seen from Figs. 19 and 20, which give the dependences of the rate of transport of zinc sulfide on the length and the square of the radius of an ampoule [71, 72]. The dependence of the rate of transport of zinc sulfide on the concentration of the transport gas, which governs the pressure of gaseous products in the ampoule, is shown in Fig. 21 [72]. The complex nature of the dependence of the transport rate on the concentration of the transport gas is due to a change in the transport mechanism. According to the results reported in [66], at low pressures in the ampoule (when the mean free path of molecules is larger than or comparable with the dimensions of the ampoule), the rate of transport is governed by the velocity of the heterogeneous reaction. This is represented in Fig. 21 by a region of rapid rise of the rate of transport with increasing pressure. At pressures $P \geq$ $100 \, N/m^2 \, (10^{-3} \, atm)$, diffusion processes are exerting an increasing influence on the transport of matter and the limits of the diffusion region extend approximately up to pressures of $300 \, kN/m^2$ (3 atm). In this range of pressures, the rate of transport does not depend appreciably on the transport gas concentration. When the pressure is increased still further, the rate of transport increases again because, in addition to diffusion, convective processes begin to play an important role.

The correct selection of the rate of transport plays a very important role in the growth of crystals by the gas-transport reaction method. When the rate of transport is too high, the supersaturation in the crystallization zone reaches a value at which many new nuclei appear and a polycrystalline ingot is formed. At too low rates of transport, the dimensions of the grown crystals are small and the time necessary to prepare them is quite long.

The transport reactions of zinc sulfide have been used as a standard in an investigation of the growth of chalcogenide crystals [68]. It was established that in the presence of chlorine in an ampoule no transport took place, that the introduction of bromine produced a small yield, and that only iodine gave satisfactory result.

These experiments were carried out in quartz ampoules 150 mm long and 8 mm in diameter, which had been previously outgassed in vacuum at 800°C. The charge weighed 3 g. When the concentration of iodine was $1 \, mg/cm^3$, the growth of crystals was slow, but when this concentration was increased to $5 \, mg/cm^3$, good results were obtained. The temperatures of the hot and cold zones were varied within wide limits but it was established that to obtain good-quality crystals the cold-zone temperature should not fall below 700°C. The following rules for the successful growth of crystals by the gas-transport reaction method were formulated in [68]:

1. The rate of transport should not be higher than the rate of growth of crystals which act as seeds.
2. The phenomenon of polymorphism should be taken into account in the selection of the crystallization temperature.
3. The crystallization zone should not be too small in order to prevent the formation of polycrystalline aggregates.
4. The temperature distribution in the crystallization zone should be uniform.
5. In those cases when the transport takes place by convection, crystals with well-developed faces are obtained.
6. A smaller temperature difference is sufficient when ampoules of large diameter are used.

The gas-transport reaction method has been used successfully in the preparation of crystals of many ternary chalcogenides of the $A^{II}B_2^{III}C_4^{VI}$ type. Table 15, which is based on the data published in [63, 68, 73-75], lists ternary chalcogenides and the conditions under which they have been prepared. It has been established [73] that ternary chalcogenides can be grown using iodine, HCl, SnI_4, and other gases. The starting material is usually prepared in the form of a mixture of binary chalcogenide powders, which is subjected to homogenizing annealing.

TABLE 15. Conditions of Growth of Ternary Chalcogenides

Compound	t_{hot}, °C	t_{cold}, °C	Amount of iodine, mg/cm³
$ZnGa_2S_4$	1100	1000	5
$CdGa_2S_4$	650	600	5
$HgGa_2S_4$	730	580	5
$ZnIn_2S_4$	1000—750	700	5—6
$CdIn_2S_4$	1000—850	750—600	5—6
$HgIn_2S_4$	950	650	5—6
$ZnGa_2Se_4$	1100	1000	5
$CdGa_2Se_4$	900	750	5
$HgGa_2Se_4$	760	710	5
$ZnIn_2Se_4$	1000—650	700—600	5—6
$CdIn_2Se_4$	1000—650	700—600	5—6

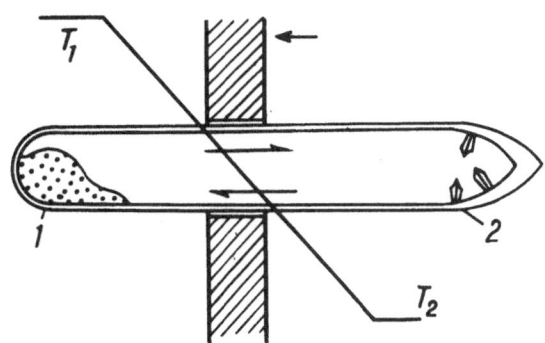

Fig. 22. Temperature distribution in a furnace during a gas-transport reaction [75]: 1) transported substance; 2) growing crystals.

Crystals are usually grown in two-section furnaces, which makes it possible to establish a temperature gradient. Figure 22 shows the temperature distribution in a furnace used to grow crystals of the ternary chalcogenides $ZnIn_2S_4$ and $CdIn_2S_4$ [75].

The gas-transport reactions taking place in the ampoule shown in Fig. 22 can be written as follows:

$$ZnS + In_2S_3 + 4I_2 \rightleftarrows ZnI_2 + 2InI_3 + 2S_2,$$
$$ZnI_2 + 2InI_3 + 2S_2 \rightleftarrows ZnIn_2S_4 + 4I_2.$$

It has been established that gas-transport reactions can be used to prepare doped crystals. In one investigation [75], the doping was achieved by the introduction of copper sulfide into an ampoule in proportions of $2 \cdot 10^{-3}$ g-atom of copper per 1 g-atom of zinc or cadmium. The concentration of copper and iodine were determined by chemical analysis. The following results were obtained:

	$\dfrac{\text{g-atom I}}{\text{g-atom S}}$	$\dfrac{\text{g-atom Cu}}{\text{g-atom Zn (Cd)}}$
$ZnIn_2S_4$	0.00067	0.00003
$ZnIn_2S_4$: Cu	0.00025	0.0019
$CdIn_2S_4$	0.00046	0.00002
$CdIn_2S_4$: Cu	0.00036	0.0019

The concentration of iodine in crystals of ternary chalcogenides can be determined by means of radioactive I^{131} [74]. It has been found that the concentration of iodine in crystals is of the order of $10^{-2}\%$ (by weight).

In some investigations, it has been found that the doping of ternary chalcogenides with copper or tellurium has improved the habit and produced larger crystals [63, 75].

Depending on the composition of the original mixture, gas-transport reactions can be used to prepare crystals of various compositions. Thus, the preparation of $Zn_3In_2S_6$ crystals, corresponding to the $3ZnS + In_2S_3$ composition of the original mixture, has been reported in [76].

There have been no reports of the preparation of ternary compounds containing oxygen or tellurium by the gas-transport reaction method. A layer of the spinel $MgAl_2O_4$ has been prepared on the surface of sapphire in an atmosphere of hydrogen and in the presence of MgO vapor at 1500-1900°C [77]. However, this layer was most probably produced not by a gas-transport reaction but by the sublimation of MgO followed by its reaction with Al_2O_3.

It should be possible to prepare ternary tellurides by the gas-transport reaction method because many binary tellurides have already been prepared by this method.

In each case the transport gas must be selected on the basis of the heats of the heterogeneous reaction. By way of example, we can quote an investigation [78] in which the use of iodine as the transport gas for the growth of Cu_2GeSe_3 crystals has failed to produce satisfactory results. However, it has been reported in the same paper that iodine can be used as the transport gas for solid solutions of the Cu_2GeSe_3–GaAs system. This apparently paradoxical result can be explained by the fact that, in the formation of solid solutions, the quantity $\Delta G = \Delta H - T\Delta S = -RT \ln K$ for heterogeneous reactions of various components of solid solutions with iodine depends on these components and this results in a decrease of the deviation from equilibrium and an increase in the amount of transported matter.

Practically no information is available on the preparation of ternary nitrides, phosphides, arsenides, and antimonides by gas-transport reactions.

It is reported in [66] that gas-transport reactions of binary nitrides are very slow because of the high stability of nitrogen molecules and low heterogeneous reaction rates. In this case, the rate of transport is governed only by the rate of the reaction and crystals grow very slowly. The theory of the growth of crystals from the gaseous phase, under restrictive conditions imposed by surface reactions, is discussed in [80].

The preparation of the ternary phosphide $ZnSiP_2$ is described in [81]. Ampoules, 120 mm long and 30 mm in diameter, were used to grow crystals. Iodine was used as the transport gas. The process was carried out at 1100°C using a temperature difference up to 40 deg; the duration of the process was 15-20 h. In the majority of cases, red needle-like crystals up to 10 mm in length, elongated along the [111] directions, were obtained. When the conditions were altered, plate-like crystals, which measured up to $6 \times 1.5 \times 0.3$ mm, were obtained. Doping made it possible to prepare p- and n-type $ZnSiP_2$ crystals.

No methods for the growth of ternary arsenides or antimonides by gas-transport reactions have yet been described. Our experiments on the growth of $CdSnAs_2$ crystals, as well as some ternary antimonides (using various transport gases), were not successful.

Thus, the problem whether gas-transport reactions can be used to grow crystals of many ternary compounds has not yet been solved. However, there is no doubt that this method for the preparation of complex crystals is promising.

CHAPTER IV

TERNARY TWO-CATION COMPOUNDS
WITHOUT DEFECTS

$A^I B^{III} C_2^{VI}$ Compounds

Most of the compounds of this type have a lattice close to that of chalcopyrite $CuFeS_2$, which is regarded as the typical representative of this group of ternary compounds, because iron in this compound is assumed to be trivalent. Semiconducting properties of natural chalcopyrite, particularly rectification at a point contact, were observed for the first time by Losev [112, 113]. More detailed information on the rectification at a point contact, as well as data on the transparency of chalcopyrite crystals at wavelengths longer than $2\,\mu$, were reported in [83, 84]. Semiconducting properties of synthetic chalcopyrite were investigated first by Boltaks and Tarnovskii [114]. Other workers [115] have established that the melting point of chalcopyrite is 950°C.

Shtrum [110] reports the results of an investigation of an analog of chalcopyrite, in which copper is replaced with silver and sulfur is replaced with tellurium. According to Zhuze, Sergeeva, and Shtrum [109], the room-temperature electron mobility in $AgFeTe_2$ exceeds $2000\,cm^2 \cdot V^{-1} \cdot sec^{-1}$. However, a detailed physicochemical investigation of substances, corresponding to the formula $AgFeTe_2$, showed [111] that they are alloys of two binary compounds.

TABLE 16. Lattice Parameters and Densities of Some
$A^I B^{III} C_2^{VI}$ Compounds

Compound	a	c	c/a	x	Density, g/cm³	
					x-ray	pycno-metric
$CuAlS_2$	5.31_2	10.4_2	1.96_1	0.27	3.47	3.45
$CuAlSe_2$	5.60_6	10.90	1.94_5	0.26	4.70	4.69
$CuAlTe_2$	5.96_4	11.7_8	1.97_5	0.25	5.50	5.47
$CuGaS_2$	5.34_9	10.4_7	1.95_8	0.25	4.35	4.29
$CuGaSe_2$	5.60_7	10.9_9	1.96_0	0.25	5.56	5.45
$CuGaTe_2$	5.99_4	11.9_1	1.98_7	0.25	5.99	5.87
$CuInS_2$	5.51_7	11.0_0	2.00_5	0.20	4.75	4.71
$CuInSe_2$	5.77_3	11.5_5	2.00_1	0.22	5.77	5.65
$CuInTe_2$	6.16_7	12.3_4	2.00_0	0.225	6.10	6.00
$CuTlS_2$	5.58_0	11.1_7	2.00_1	0.19	6.32	6.07
$CuTlSe_2$	5.83_2	11.6_3	1.99_5	0.23	7.11	7.08
$AgAlS_2$	5.69_5	10.2_6	1.80_2	0.30	3.94	3.93
$AgAlSe_2$	5.95_6	10.7_5	1.80_5	0.27	5.07	4.99
$AgAlTe_2$	6.29_6	11.8_3	1.87_8	0.26	6.18	6.15
$AgGaS_2$	5.74_3	10.2_6	1.78_6	0.28	4.72	4.58
$AgGaSe_2$	5.97_3	10.8_8	1.82_3	0.27	5.84	5.71
$AgGaTe_2$	6.28_3	11.9_4	1.89_7	0.26	6.05	5.96
$AgInS_2$	5.81_6	11.1_7	1.92_0	0.25	5.00	4.97
$AgInS_2$ (wurtzite)	4.12_1	6.67_4	1.61_8	—	4.83	—
$AgInSe_2$	6.09_0	11.6_7	1.91_6	0.25	5.81	5.80
$AgInTe_2$	6.40_6	12.5_6	1.96_2	0.25	6.12	6.08

TABLE 17. Lattice Parameters and Densities of
Oxygen-Containing $A^I B^{III} C_2^{VI}$ Compounds [31]

Compound	Hexagonal cell, nm (Å)				Rhombohedral cell, nm (Å)			Density, g/cm³	
	a	c	c/a	Z	a	$\alpha°$	Z	x-ray	pycno-metric
$CuAlO_2$	0.286_4 (2.86_4)	1.69_5 (16.9_5)	5.91	3	0.588_4 (5.88_4)	28.1	1	5.10_4	4.89_7
$CuGaO_2$	0.302_7 (3.02_7)	1.70_9 (17.0_9)	5.65	3	0.595_4 (5.95_4)	29.4	1	6.03_4	5.75_3

Hahn et al. [31] were the first to prepare a large group of normal-valence $A^I B^{III} C_2^{VI}$ compounds, with copper and silver from the IB subgroup, with aluminum, gallium, indium, or thallium from the IIIB subgroup, and with sulfur, selenium, or tellurium from the VIB subgroup. An x-ray diffraction investigation of samples obtained by Hahn et al. showed that almost all the synthesized compounds had the chalcopyrite structure. Table 16 lists the lattice parameters as well as the values of the x-ray and pycnometric density of twenty of $A^I B^{III} C_2^{VI}$ compounds prepared by Hahn et al. [31]. Only two compounds containing thallium could be prepared: $CuTlS_2$ and $CuTlSe_2$. It was found that $AgInS_2$ existed in two crystallographic modifications: chalcopyrite and wurtzite.

Attempts to synthesize $A^I B^{III} C_2^{VI}$ compounds containing oxygen produced only two materials [82]: $CuAlO_2$ and $CuGaO_2$. These compounds can exist in two crystallographic modifications: hexagonal and rhombohedral. The values of the lattice parameters of these compounds and their densities are listed in Table 17.

Theoretical calculations of the forbidden band width, carried out using the electronegativity concept [85], have shown that $A^I B^{III} C_2^{VI}$ ternary compounds (with the exception of those containing oxygen) should be semiconductors with a forbidden band width ranging from 2.5 eV ($CuAlS_2$) to 0.1 eV ($AgTlTe_2$). The electrical properties of $CuAlO_2$ and $CuGaO_2$ have not yet been investigated.

Batsanov [86] calculated the forbidden band widths from the electronegativities of the components and he suggested a formula for estimating the forbidden band of a compound from the forbidden bands of the components, the average value of the principal quantum number of valence electrons, and the average difference of the electronegativities of the components. The values of the forbidden band widths, obtained by Batsanov for some $A^I B^{III} C_2^{VI}$ compounds, are in satisfactory agreement with the experimental values.

Methods for synthesizing $A^I B^{III} C_2^{VI}$ compounds can be divided into two groups. One group includes the synthesis of ternary compounds, by semiconductor metallurgy methods, from binary components prepared by standard chemical techniques [31, 82]. The other group includes synthesis by the fusion of the constituent elements in evacuated and sealed quartz ampoules, which is preceded by heating in several stages, corresponding to temperatures close to the melting points of the components [93], and followed by slow cooling or the slow withdrawal of an ampoule from a furnace [94]. Slow withdrawal at an approximate rate of 3 cm/h has yielded polycrystalline samples with large-block structure [94].

Compounds containing aluminum are synthesized in closed Alundum crucibles, because at high temperatures aluminum reacts with quartz to form silicates. Alundum crucibles are placed in evacuated and sealed quartz ampoules.

TABLE 18. Parameters of Some $A^I B^{III} C_2^{VI}$ Compounds [93]

Compound	t_{mp}, °C	ΔE, eV	α, μV/deg	a, nm (Å)	c/a
$CuAlSe_2$	1000	1.10	—	—	—
$CuGaSe_2$	861	0.96	+4.6	0.600 (6.00)	1.97
$CuInSe_2$	980	0.86	+613	0.576 (5.76)	2.03
$AgAlSe_2$	950	0.70	—	—	—
$AgGaSe_2$	883	0.66	—	0.596 (5.96)	1.91
$AgInSe_2$	780	0.62	—226	0.594 (5.94)	2.18
$CuAlTe_2$	890	0.88	+14	0.593 (5.93)	2.01
$CuGaTe_2$	872	0.82	+235	0.609 (6.09)	1.81
$CuInTe_2$	791	0.67	+237	0.617 (6.17)	1.98
$AgAlTe_2$	729	0.56	+321	0.624 (6.24)	1.92
$AgGaTe_2$	714	0.52	+346	0.650 (6.50)	1.61
$AgInTe_2$	692	0.52	+298	0.646 (6.46)	1.87

Austin, Goodman, and Pengelly [89] reported that ingots of $CuInSe_2$ and $AgInTe_2$ compounds synthesized by them had cracks and these cracks could not be removed by zone melting or directional recrystallization. They attributed this effect to the anisotropy of thermal expansion. However, according to the results obtained by one of the present authors and his colleagues [100, 101], crack-free polycrystalline samples of these compounds can be obtained by reducing considerably the rate of cooling of the melt and by annealing for 24-48 h at a temperature a little lower than the solidus temperature. Similar results were obtained also for alloys of some binary compounds [145].

Some investigators have used zone recrystallization to purify $A^I B^{III} C_2^{VI}$ ternary compounds [89].* However, zone recrystallization or Bridgman's directional crystallization has failed to produce single crystals of these compounds. A single-crystal ingot has been obtained by zone melting only for one of the alloys of the $CuInTe_2 - CdTe$ system [108].

Our own attempts to prepare single crystals of $CuInTe_2$ by the Czochralski pulling method under an inert gas pressure were unsuccessful. In spite of the relatively low vapor pressure of tellurium and a considerable pressure of the inert gas, tellurium evaporated rapidly from the melt and was deposited on the walls of the container.

There have been many investigations of the electrical, optical, thermal, and other properties of $A^I B^{III} C_2^{VI}$ compounds. Goodman and Douglas [87] have reported a large inverse voltage (up to 150 V) for crystals of $CuInSe_2$ and $AgInSe_2$. Measurements carried out on polycrystalline samples of $CuInSe_2$ have indicated an electrical resistivity of 0.1 $\Omega \cdot$ cm, a Hall mobility of about 300 $cm^2 \cdot V^{-1} \cdot sec^{-1}$, and an optical value of the forbidden band width of about 0.9 eV. The carrier mobility should be higher in single crystals of this compound, but attempts to prepare single crystals have not been successful.

The results of measurements of the optical value of the forbidden band width ΔE (eV) of five $A^I B^{III} C_2^{VI}$ compounds, reported in [88], are given on the following page.

---

*See also V. M. Koshkin, Dissertation for Candidate's Degree [in Russian], Khar'kov State University (1964).

$CuInS_2$ 1.2
$CuInSe_2$ 0.92
$CuInTe_2$ 0.95
$AgInSe_2$ 1.18
$AgInTe_2$ 0.96

Investigations of $CuInSe_2$ samples [89], which exhibit a considerable point–contact rectification [90], have established their twinning tendency. Moreover, thermal disordering at temperatures above 700°C, similar to that observed in $CuFeS_2$ [91], has been reported for this compound. Austin, Goodman, and Pengelly [89] noted that the zone–melted ingots of $CuInSe_2$ exhibit some variation of the lattice parameters along the ingot length. This indicates a range of solubility of selenium and copper, but the results of optical measurements carried out on samples, prepared from different parts of the zone–melted ingot, have yielded the same values.

Investigations of Palatnik and his colleagues [116, 125] of quasibinary $CuInSe_2 - In_2Se_3$ and $CuGaSe_2 - Ga_2Se_3$ alloys have indicated that binary compounds are soluble in ternary materials and that the optical value of the forbidden band width depends on the alloy composition. Similar results were published in [96] for $Cu_2Te - In_2Te_3$ quasibinary alloy subsystem of the copper– indium–tellurium ternary system; the $CuInTe_2$ compound is in the middle of this subsystem.

Investigations of the temperature dependence of the forbidden band width, deduced from the absorption and photoconductivity spectra, are described in [101]. A very important result follows from these investigations, which has not yet been explained satisfactorily: compounds of the investigated type include samples whose average temperature coefficient of the forbidden band width is either zero or positive in the temperature range 80-300°K. This indicates some special features in the band structure of the investigated compounds.

A contactless method [92] has been used to determine the temperature dependences of the electrical conductivity of twelve $A^I B^{III} C_2^{VI}$ compounds [93]. All the investigated compounds, with the exception of $AgInSe_2$, have p-type conduction. The same paper [93] gives the values of the melting points, microhardness, forbidden band widths, thermoelectric powers, and lattice parameters of these twelve compounds (Table 18).

Table 18 shows that the thermoelectrical powers of some of these compounds are high. Generally speaking, such high powers can be explained, as pointed out in [117], by a distribution of the excess components (or binary compounds, in the case of a deficiency of one of the components) along grain boundaries, which turns a crystal into a battery of thermoelements, as reported also in [94].

TABLE 19. Melting Points t_{mp}, Microhardness H, Forbidden
Band Widths ∆E, Electrical Conductivities σ,
Thermoelectric Powers $α$, and Linear Expansion
Coefficients $α_L$ of Some $A^I B^{III} C_2^{VI}$ Compounds [94, 134]

Compound	t_{mp}, °C	H ± 20, $\frac{kg}{mm^2}$	∆E, eV	σ, $Ω^{-1} \cdot cm^{-1}$		$α$, µV/deg		$α_L \cdot 10^6$, deg^{-1}
				before annealing	after annealing	before annealing	after annealing	
$CuGaSe_2$	1040	430	1.63	0.02	0.06	+75	+40	3.4
$CuGaTe_2$	870	360	1.0	11	6	+270	+340	11.2
$CuInSe_2$	990	260	1.07	0.14	0.04	+480	+640	7.1
$CuInTe_2$	780	210	0.95	100	65	+140	+260	8.6
$CuTlSe_2$	405	90	—	6000	1000	+10	−5	—
$CuTlTe_2$	375	100	—	2800	—	+80	—	—
$AgGaSe_2$	850	450	1.66	8.8	0.12	−70	+90	—
$AgGaTe_2$	720	180	1.1	$1.7 \cdot 10^{-4}$	$1.7 \cdot 10^{-5}$	+700	+950	4.2
$AgInSe_2$	773	230	1.18	0.48	$5 \cdot 10^{-5}$	−140	−370	—
$AgInTe_2$	680	190	0.93	0.78	10^{-5}	−70	−100	5.5
$AgTlSe_2$	328	120	0.72	10^{-5}	—	+800	—	—
$AgTlTe_2$	290	140	—	410	1800	+60	+10	12.8

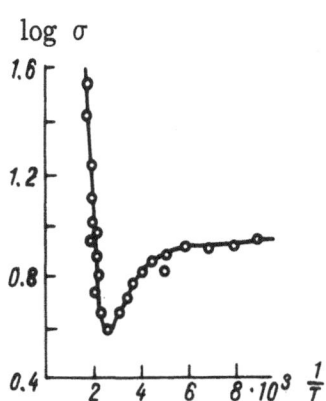

Fig. 23. Temperature dependence of the electrical conductivity of $CuInSe_2$.

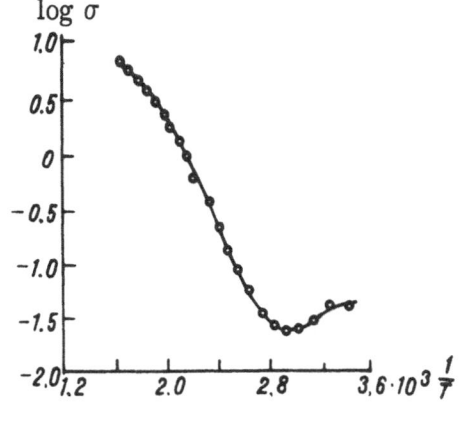

Fig. 24. Temperature dependence of the electrical conductivity of $AgInTe_2$.

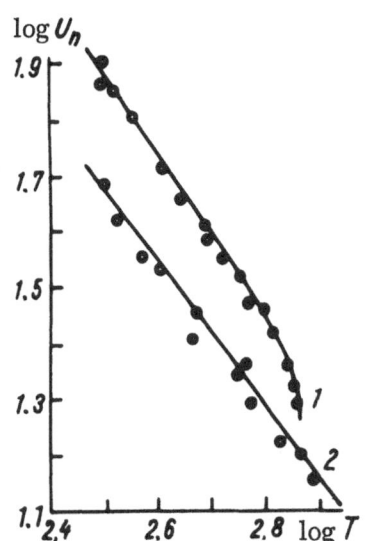

Fig. 25. Temperature dependences of the carrier mobility: 1) $CuInTe_2$; 2) $CuGaTe_2$.

Zhuze, Sergeeva, and Shtrum [94] have investigated in detail the preparation techniques and have measured the melting points, microhardness, electrical conductivities, thermoelectric powers, and thermal expansion coefficients of a large group of $A^IB^{III}C_2^{VI}$ compounds. They have also investigated the influence of prolonged annealing on the electrical conductivity and thermoelectric power (Table 19).

The temperature dependences of the electrical conductivity reported in [94] are typical of extrinsic semiconductors (Figs. 23 and 24). It has also been found that the Hall mobility of carriers in $CuInTe_2$ is proportional to $T^{-1.6}$ and the mobility in $AgInTe_2$ is proportional to $T^{-1.3}$; this may be attributed to the scattering of carriers by the nonpolar acoustical vibrations. The temperature dependences of the carrier (hole) mobility in $CuInTe_2$ and $CuGaTe_2$ are shown in Fig. 25.

Zhuze, Sergeeva, and Shtrum [94] failed to alter the sign of conduction in ternary $A^IB^{III}C_2^{VI}$ compounds by deviation from stoichiometry. In this respect, such compounds behave like binary $A^{III}B^V$ semiconductors. The similarity of the linear expansion coefficients of $A^IB^{III}C_2^{VI}$ compounds and of elemental semiconductors of the fourth group [106] and binary $A^{III}B^V$ compounds [118] indicates that the nature of the forces and the values of the atomic interaction energies are similar in all these substances.

Practical applications of ternary $A^IB^{III}C_2^{VI}$ compounds are discussed by Apple [119], who investigated the influence of $CuGaS_2$ and $AgGaS_2$ admixtures on the luminescence properties of ZnS phosphors. He found that these ternary compounds are soluble in zinc sulfide and that lattice parameters vary with the concentration of the components in accordance with the Vegard law, within the limits of the existence of solid solutions (up to 33.3 mol.% $CuGaS_2$ and 5-10 mol.% $AgGaS_2$). Samples of solid solutions luminesce when excited with radiation of the 365 nm (3650 Å) wavelength, in accordance with the Stokes law; an increase of the concentration of the ternary component in the solution shifts the luminescence maximum in the direction of longer wavelengths, as indicated by the spectra shown in Fig. 26.

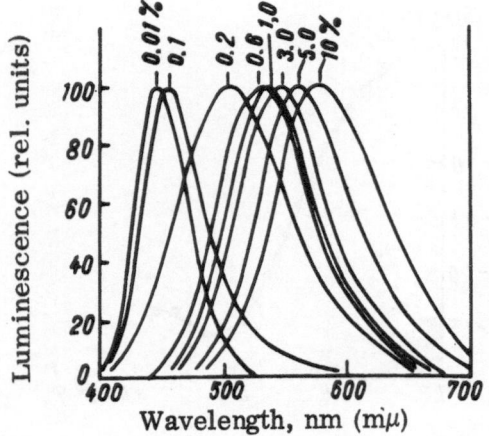

Fig. 26. Luminescence spectra of
ZnS−AgGaS$_2$ solid solutions with various amounts of AgGaS$_2$ (indicated
alongside the curves).

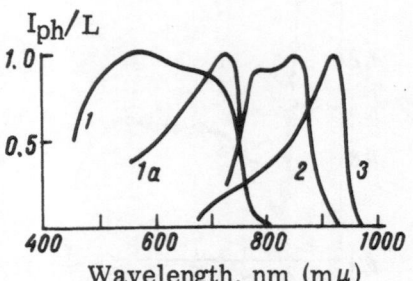

Fig. 27. Photoconductivity spectra: 1, 1a) CuGaSe$_2$; 2) CuTlTe$_2$;
3) CuInSe$_2$.

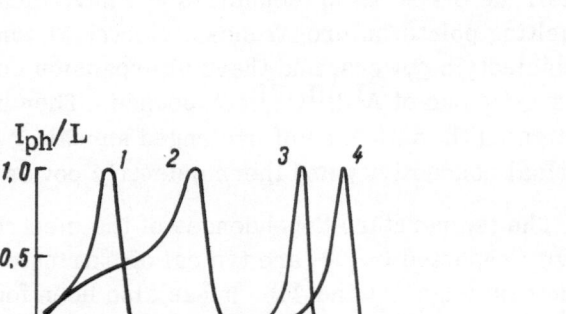

Fig. 28. Photoconductivity
spectra: 1) AgGaSe$_2$; 2)
AgGaTe$_2$; 3) AgInSe$_2$; 4)
AgInTe$_2$.

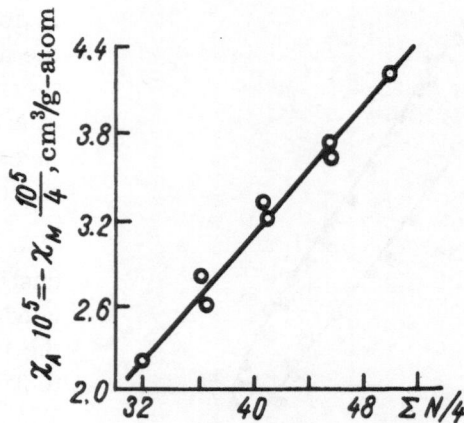

Fig. 29. Dependence of the average atomic magnetic susceptibility
of some A^{I}B^{III}C$_2^{VI}$ compounds on
the average atomic number.

TABLE 20. Melting Points t$_{mp}$, Microhardness H,
Thermal Conductivities $\varkappa$, and Linear Expansion
Coéfficients α_L of Some A^{I}B^{III}C$_2^{VI}$
Compounds [95]

Compound	t$_{mp}$, °C	H		$\varkappa$		$\alpha_L \cdot 10^6$ deg^{-1}
		MN/m^2	kg/mm^2	W·m^{-1}· deg^{-1}	cal·cm^{-1}· sec^{-1}·deg^{-1} ·10^3	
CuGaSe$_2$	1040	4267	435	4.18	10	5.4
CuGaTe$_2$	870	3404	347	2.7	6.5	6.9
CuInSe$_2$	990	2462	251	2.9	7.0	6.6
CuInTe$_2$	778	1864	190	2.5	5.9	7.1
AgGaSe$_2$	840	3041	310	2.65	6.35	—
AgGaTe$_2$	705	2550	260	1.85	4.4	1.9
AgInSe$_2$	775	1834	187	1.93	4.6	—
AgInTe$_2$	680	1864	190	1.55	3.7	4.3

TABLE 21. Densities ρ, Velocities of Longitudinal
Ultrasound v, Young's Moduli E, and Characteristic
Debye Temperatures θ of Some $A^IB^{III}C_2^{VI}$
Compounds [95]

Compound	ρ, g/cm³	V, m/sec	E		θ, °K
			GN/m²	Tdyn/cm²	
CuGaSe₂	5.27	—	—	—	280
CuGaTe₂	5.47	3240	60.4	0.604	183
CuInSe₂	5.89	3430	68.8	0.688	218
CuInTe₂	6.38	3240	64	0.640	167

The optical and photoelectric properties of ternary $A^IB^{III}C_2^{VI}$ compounds have not yet been investigated sufficiently thoroughly [94, 101]. The published data are available mainly on the absorption spectra at room temperature [88, 91], which have been used to estimate the forbidden band width. The room-temperature photoconductivity spectra, reported in [94] for some ternary compounds of the $A^IB^{III}C_2^{VI}$ type, are shown in Fig. 27 (compounds of copper) and in Fig. 28 (compounds of silver).

The magnetic properties of $A^IB^{III}C_2^{VI}$ compounds were investigated by one of the present authors and his colleagues [102]. Measurements have been made of the magnetic susceptibility of some of these compounds and the purpose of these measurements was to study the influence of the degree of distortion of valence angles betweeen directional bonds (the degree of tetragonality) on the physical properties. The results have shown that the investigated compounds are diamagnetic and their susceptibility is independent of the ratio c/a of the lattice parameters but is proportional to the average atomic number of the compound (Fig. 29), in agreement with the current theory of the nature of diamagnetism.

In his dissertation, Koshkin pointed out that some physical properties, particularly the melting point, should be sensitive to the degree of tetragonality of the unit cells of these compounds. However, this does not apply to the magnetic properties.

The thermal properties of $A^IB^{III}C_2^{VI}$ compounds have not yet been investigated in sufficient detail. The melting points have been reported in a number of papers [39, 93, 94, 99, 109]. The values of the thermal conductivity of some compounds can be found in [96, 99]. The thermal expansion coefficients and the Debye temperatures have also been determined [94, 96, 99, 105, 120]. Since compounds of the $A^IB^{III}C_2^{VI}$ type have tetragonal lattices, it is quite likely that the thermal expansion coefficients of these compounds should have different values along different crystallographic directions, but the anisotropy of the thermal expansion has not yet been confirmed because of the lack of success in growing single crystals. The results of measurements of the melting points, thermal expansion coefficients, and thermal conductivities are given in Table 20.

The temperature dependence of the thermal expansion coefficient of CuInTe₂ in the range 20-340°K, reported in [108], is of great theoretical interest. It has been found that below about 43°K, the thermal expansion coefficient of this compound becomes negative like the coefficients of other diamond-like (with the exception of diamond itself) and zinc-blende-type substances [104, 106, 121, 122, 123]. The results reported in [103] confirm that the complication of the structure along the diamond—sphalerite—chalcopyrite series does not alter basically the phonon spectrum and does not affect the horizontal part of the dispersion curve of the transverse acoustic vibrations [107]. Novikova [103] assumed that the sign of the thermal expansion coefficient is affected by the long-range forces associated with the polarization of atoms.

The elastic properties of $A^IB^{III}C_2^{VI}$ compounds have also been investigated [95]. Table 21 lists the values of the densities, velocities of propagation of longitudinal ultrasonic waves, Young's moduli, and Debye temperatures taken from [95].

The thermoelectric properties of $A^IB^{III}C_2^{VI}$ compounds, which have to be known in order to find whether these materials can be used to transform heat into electrical power, have not yet been investigated in detail. This is because $A^IB^VC_2^{VI}$ and $A^IB^{VIII}C_2^{VI}$ ternary compounds as well as alloys of these compounds are attracting more attention [109, 124] because they have a promising combination of thermoelectric parameters (thermoelectric power, electrical conductivity, and thermal conductivity).

Some scatter of the parameters (cf., for example, [94, 97]) can be seen in Tables 16-20. This is due to the different methods used to synthesize these compounds, which govern the phase composition, crystallite size, etc. Obviously, methods for the synthesis of particular ternary compounds can be selected reliably only after detailed phase diagrams are obtained.

The nature of the formation of $A^IB^{III}C_2^{VI}$ compounds has not yet been finally determined. According to Mason and O'Kane [39], these compounds are formed by peritectic reactions, but their view is not supported by examination of the phase diagrams. Published data are available only on quasibinary $CuGaSe_2-Ga_2Se_3$ and $CuInSe_2-In_2Se_3$ systems [116, 125] and on $Cu_2Te-In_2Te_3$ and $Cu_2Te_3-In_2Te$ systems [96].

It is reported in [94] that the sign of carriers in $CuInSe_2$ can be altered by heating in cadmium vapor, which can be used to produce p—n junctions in these crystals. In this connection, it is worth mentioning the work of Voitsekhovskii [98] on the range of existence of solid solutions in the $CdTe-CuInTe_2$ and $InAs-CuInTe_2$ systems, as well as the investigations of Chernyavskii [108].

It is known [241] that the covalent tetrahedral binding prevents the formation of glasses. However, it has been reported in [126, 240] that some $A^{II}B^{IV}C_2^V$ ternary compounds can exist in the glassy state. This has suggested that compounds of other types (particularly ternary sulfides and selenides) may form glassy semiconductors. Unfortunately, all our attempts to prepare glassy $A^IB^{III}C_2^{VI}$ compounds were unsuccessful.

It is difficult to indicate the possible range of practical applications of $A^IB^{III}C_2^{VI}$ compounds. The high reverse voltage (up to 400 V) of $CuInSe_2$ point-contact diodes [88, 89] is undoubtedly of interest. Since these compounds are formal analogs of $A^{II}B^{VI}$ compounds, it is likely that their applications will be similar to those of $A^{IV}B^{VI}$ compounds [119]. Measurements of the carrier mobility [94] do not support the conclusions of Austin, Goodman, and Pengelly [89] that one can hardly expect high mobilities in $A^IB^{III}C_2^{VI}$ compounds.

Recently, interest has increased in oxygen-containing compounds of the $A^IB^{III}O_2$ type, where A^I are elements of the IA subgroup (lithium, sodium, etc.), because it has been reported [128] that β-$LiGaO_2$ is a very efficient piezoelectric. The piezoelectric properties of these compounds can be explained by the specific features of their crystal structure. There have been reports of the synthesis and investigations of the structure of $NaAlO_2$ [129, 130], $LiGaO_2$ [131], as well as $NaFeO_2$ [132, 133] and $LiFeO_2$ [327], but no detailed investigations of the physicochemical and electrical properties of these compounds have yet been undertaken.

This review of the information on $A^IB^{III}C_2^{VI}$ compounds shows that the studies of these compounds and of their potential practical applications are still embryonic. The most pressing problems are the investigation of the phase diagrams of $A^I-B^{III}-C^{VI}$ systems; the development of techniques for the preparation of single crystals of compounds, as well as of their solid solutions with their own components and corresponding binary compounds; and the elucidation of the physicochemical and electrical properties of single crystals. In our opinion, it would be very interesting to investigate the following systems of quasibinary alloys belonging to the $A^I - B^{III} - C^{VI}$ ternary system: $AgInTe_2 - In_2Te_3$ [135], $AgGaTe_2 - Ga_2Te_3$ [136], $AgInSe_2 - In_2Se_3$ [137], $CuGaSe_2 - Ga_2Se_3$ [125, 138, 139], $CuInSe_2 - In_2Se_3$ [116, 125, 140], $CuGaTe_2 - Ga_2Te_3$ [141], and $CuInTe_2 - In_2Te_3$ [142].

All these systems of alloys are parts of quasibinary subsystems $A_2^I C^{VI} - B_2^{III} C_3^{VI}$, satisfying the condition of normal valence. It would be very desirable to carry out physicochemical investigations of quasibinary subsystems $A_2^I C_3^{VI} - B_2^{III} C^{VI}$, which satisfy the four-electron condition.

In conclusion, we must stress that, in spite of the scarcity of information on $A^I B^{III} C_2^V$ compounds, there are no grounds for assuming that they are unlikely to find practical applications in semiconductor technology. This view is supported by the work of Douglas and Goodman [90], who have determined the current-voltage characteristic of a $CuInSe_2$ point-contact diode and have compared it with a germanium diode. This comparison has shown that the characteristics of both substances are practically identical in the reverse direction, but in the forward direction the ternary compound has a somewhat higher resistance, which Douglas and Goodman [90] attributed to the absence of the formation stage during preparation of the diode. Of great practical interest are the investigations of the method of preparation and an electron-diffraction study of the structure of semiconducting $AgTlTe_2$ films, carried out by Imamov and Pinsker [143, 144].

$A_2^I B^{IV} C_3^{VI}$ Compounds

Ternary $A_2^I B^{IV} C_3^{VI}$ compounds were first mentioned by Goodman [10], who synthesized Cu_2SnSe_3, Cu_2SiTe_3 and Cu_2SnTe_3, and by Hahn [148]. Averkieva and Vaipolin [149] reported that $A_2^I B^{IV} C_3^{VI}$ compounds (where A^I is copper, B^{IV} is tin, and C^{VI} is sulfur, selenium, or tellurium) have the sphalerite structure. The lattice parameters, structure and "superstructure" lines in x-ray diffraction patterns, as well as microhardness and melting points of some $A_2^I B^{IV} C_3^{VI}$ compounds can be found in [150].

The problem of deviation of the experimentally obtained lattice parameters from those calculated by Pauling and Huggins [26] is considered in [151], where it is shown that, in compounds with mixed interatomic forces, the ionic component of these forces must be taken into account in a consistent manner.

The first attempts to synthesize $A_2^I B^{IV} C_3^{VI}$ compounds containing silver were announced in [149, 152]. It was found [149] that Ag_2SnSe_3 and Ag_2SnS_3 alloys have the single-phase structure and the same crystal lattice, but their structure differs from that of sphalerite. It is reported in [152] that a section of an alloy corresponding to the composition Ag_2SnTe_3 shows the presence of two phases, one of which (representing 60% of the crystal) has the sphalerite structure.

TABLE 22. Lattice Constants a, Electrical Conductivities σ, Thermoelectric Powers α, and Forbidden Band Widths of Some $A_2^I B^{IV} C_3^{VI}$ Compounds [155]

Compound	a, nm (Å)	σ, $\Omega^{-1} \cdot cm^{-1}$	α, $\mu V/deg$	ΔE, eV
Cu_2GeS_3	0.53 (5.30)	1.9	100—300	0.3
Cu_2GeSe_3	0.55 (5.55)	50	70—100	—
Cu_2GeTe_3	0.59 (5.95)	$1.4 \cdot 10^3$	10	—
Cu_2SnS_3	0.54 (5.43)	0.49	100—600	—
Cu_2SnSe_3	0.56 (5.68)	91	250	0.7 ± 0.1
Cu_2SnTe_3	0.60 (6.04)	$1.4 \cdot 10^4$	30	—

TABLE 23. Lattice Parameters and Densities
of Some $A_2^I B^{IV} C_3^{VI}$ Compounds [46]

Compound	Crystal system	Lattice parameters, nm(Å)			Density, g/cm^3	
		a	c	c/a	exper.	calc.
Cu_2SiS_3 {	Hexagonal (wurtzite)*	0.368 (3.684)	0.600(6.004)	1.641	3.81	3.89
	Tetragonal †	0.529 (5.290)	1.015(10.156)	1.920	3.63	3.89
Cu_2SiTe_3	Cubic or tetragonal	0.593 (5.93)	—	—	5.47	5.69
Cu_2GeS_3 {	Cubic *	0.532 (5.317)	—	—	4.45	4.36
	Tetragonal †	0.533 (5.326)	1.044(10.438)	1.960	4.46	4.43
Cu_2GeSe_3	do.	0.559 (5.595)	1.056(10.564)	1.960	5.57	5.63
Cu_2GeTe_3	do.	0.596 (5.956)	1.185(11.852)	1.990	5.95	6.13
Cu_2SnS_3	Cubic	0.544 (5.445)	—	—	5.02	4.69
Cu_2SnSe_3	do.	0.570 (5.696)	—	—	5.94	5.79
Cu_2SnTe_3	do.	0.605 (6.047)	—	—	6.51	6.29

*High-temperature modification.
†Low-temperature modification.

Compounds of this type are usually synthesized by the fusion of elements in evacuated and sealed quartz ampoules. Zone recrystallization can be used successfully in purification of these compounds [151]. Compounds obtained in this way [146] have, according to x-ray diffraction analysis, cubic diamond-like crystal lattices, but these lattices are typical only of the high-temperature disordered modifications (in those cases in which ordering is possible), whereas at low temperatures they have a structure similar to that of chalcopyrite.

Palatnik et al. [24] consider that the data on the tetragonality of some $A_2^I B^{IV} C_3^{VI}$ compounds investigated by them cannot be regarded as final because the excess of group IV elements in these compounds, reported in [153], reduces the degree of tetragonality of the lattice.

It is interesting to compare the lattice parameters of Cu_2GeSe_3 with those of other compounds of the isoelectronic germanium series [154]. It is found that this ternary compound satisfies the rule suggested by Goldschmidt [147] for isoelectronic series.

The electrical properties of $A_2^I B^{IV} C_3^{VI}$ compounds were reported first by Palatnik et al. [151, 155]. The lattice parameters, electrical conductivities, thermoelectric powers, and forbidden band widths of some of these compounds are listed in Table 22 [155]. All these compounds exhibit p-type conduction. It is reported in [151] that the sign of carriers in these compounds can be altered by the departure of the composition from stoichiometry, but our own investigations [328] have not confirmed this conclusion: the sign of carriers is unaffected by large deviations (up to 10%) of the concentration of one of the components from the stoichiometric proportions.

A large group of $A_2^I B^{IV} C_3^{VI}$ compounds was studied by a team of French investigators [46]. These compounds have been synthesized by the fusion of elements with about 1% excess of sulfur, selenium, or tellurium, compared with the stoichiometric amount. Thermal analysis has shown that Cu_2SiS_3 has two enantiotropic forms with a transition point near 840°C. Quenching in cold water from temperatures above 840°C produces a wurtzite-type structure in Cu_2SiS_3, but after annealing at lower temperatures the lattice is tetragonal. It is reported in [46] that attempts to synthesize Cu_2SiSe_3 have not been successful.

Cu_2SiTe_3 is difficult to synthesize because it dissociates at low temperatures to form CuTe. It has not been possible to establish whether the lattice of Cu_2SiTe_3 is cubic (of the sphalerite type) or tetragonal. Cu_2GeS_3, like Cu_2SiS_3, exhibits a phase transition at 670°C: the

TABLE 24. Forbidden Band Widths
of Some $A_2^I B^{IV} C_3^{VI}$ Compounds [157]

Compound	ΔE_{elec}, eV	$\Delta E_{photoelec}$, eV	
		293 °K	77 °K
Cu_2SnS_3	—	0.91 ± 0.01	0.93 ± 0.01
Cu_2SnSe_3	0.5 ± 0.05	0.66 ± 0.05	0.78 ± 0.02
Cu_2GeSe_3	0.6 ± 0.1	0.94 ± 0.05	0.77 ± 0.05
Ag_2SnSe_3	0.7 ± 0.1	0.81 ± 0.05	0.84 ± 0.05
Ag_2SnS_3	0.5 ± 0.1	—	—
Ag_2GeSe_3	0.9 ± 0.1	—	0.91 ± 0.05

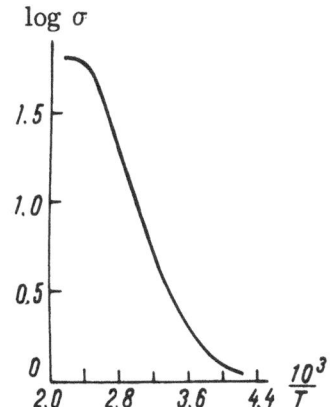

Fig. 30. Temperature dependence of the electrical conductivity of Ag_2SnS_3.

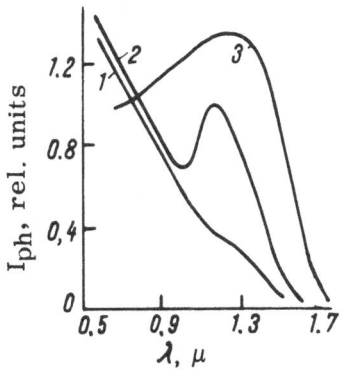

Fig. 31. Photoconductivity spectra: 1) Ag_2GeSe_3 (20°C); 2) Ag_2GeSe_3 (−196°C); 3) Ag_2SnSe_3 (20°C).

high-temperature modification has a cubic lattice and the low-temperature modification is tetragonal. The lattice parameters reported in [46] are in good agreement with Hahn's values [148] and the lattice parameters of compounds of tin agree with the results of Palatnik et al. [146]. Some of the results reported in [46] are given in Table 23.

A more detailed investigation of the structure of some $A_2^I B^{IV} C_3^{VI}$ compounds, reported in [156], has confirmed the results of the French workers [46]. However, in our opinion, the problem of the structure of these compounds cannot yet be regarded as finally solved. This remark is prompted by the presence of unidentified "superstructure" lines in the x-ray diffraction patterns and by the fractional number of formula units of the compound ($^4/_3$) in each unit cell. It is very likely that the unit cell of many of $A_2^I B^{IV} C_3^{VI}$ compounds is in the form of a cube (or tetrahedron) which is repeated three times along some direction.

The photoelectric properties as well as the temperature dependences of the electrical conductivities of some $A_2^I B^{IV} C_3^{VI}$ compounds were determined by Kharakhorin and Petrov [157]. They found the forbidden band widths of six compounds from the photoconductivity spectra, photo-emf, and photomagnetic effect. They also determined the temperature dependences of the electrical conductivity and Hall effect. The temperature dependence of the electrical conductivity of Ag_2SnS_3 [157] is given in Fig. 30. The photoconductivity spectra of three $A_2^I B^{IV} C_3^{VI}$ compounds are given in Fig. 31. The values of the forbidden band widths of the compounds investigated in [157] are listed in Table 24.

TABLE 25. Melting Points and Microhardness
of Some $A_2^I B^{IV} C_3^{VI}$ compounds

Compound	Melting point, °C				H [158]	
	[46]	[158]	[153]	[328]	MN/m²	kg/mm²
Cu_2SiS_3	925	—	—	—	—	—
Cu_2GeS_3	933	955	948	956	4552	464
Cu_2GeSe_3	760	788	770	783	3836	391
Cu_2GeTe_3	595	492	--	—	2894	295
Cu_2SnS_3	838	855	845	874	2773	283
Cu_2SnSe_3	685	697	696	698	2511	256
Cu_2SnTe_3	410	411	—	—	1972	201

TABLE 26. Linear Expansion Coefficients α_L, Velocities
of Ultrasonic Waves v, Young's Moduli E, Characteristic
Debye Temperatures θ, and Specific Heats c
of Some $A_2^I B^{IV} C_3^{VI}$ Compounds [328]

Compound	$\alpha_L \cdot 10^6$, deg^{-1}	v, km/sec	E		θ, °K	c	
			GN/m²	Tdyn/cm²		J·g^{-1} ·deg^{-1}	cal·g^{-1} ·deg^{-1}
Cu_2GeS_3	7.2	5.66	14	0.140	254	0.51	0.121
Cu_2SnS_3	7.8	4.95	11.6	0.116	214	0.44	0.105
Cu_2GeSe_3	8.4	4.02	9.1	0.091	168	0.34	0.082
Cu_2SnSe_3	8.9	3.61	7.5	0.075	148	0.31	0.074

TABLE 27. Electrical Conductivities σ, Hall Coefficients R,
Carrier Densities n, and Mobilities μ
of Some $A_2^I B^{IV} C_3^{VI}$ Compounds [328]

Compound	σ, $\Omega^{-1} \cdot cm^{-1}$	R, cm³/C	$n \cdot 10^{-17}$, cm^{-3}	μ cm²·V^{-1}·sec^{-1}
Cu_2GeS_3	17.3	28.5	3.0	360
Cu_2SnS_3	39.6	10.2	6.1	405
Cu_2GeSe_3	5.71	57.0	1.5	283
Cu_2SnSe_3	71.0	12.3	5.1	870

The melting points of $A_2^I B^{IV} C_3^{VI}$ compounds are given in [46, 158, 159, 328]. The results of measurements of the melting points, together with the values of microhardness, are listed in Table 25; they show that the melting points obtained by different authors are in good agreement.

It has been established [46, 158], that $A_2^I B^{IV} C_3^{VI}$ compounds containing copper and tin have the sphalerite structure and those containing copper and germanium have tetragonal lattices.

The results of systematic studies of the thermal and elastic properties of a group of $A_2^I B^{IV} C_3^{VI}$ compounds were published in [160, 161, 328]. The properties investigated included the thermal expansion, microhardness, and velocity of longitudinal ultrasonic waves. The results were used to calculate the characteristic Debye temperatures, Young's moduli, specific heats, and mean free paths of phonons. Some of these results are presented in Table 26.

To determine the quality of the prepared samples and the possibilities of practical applications, measurements have been made of the sign, density, and mobility of carriers [328]

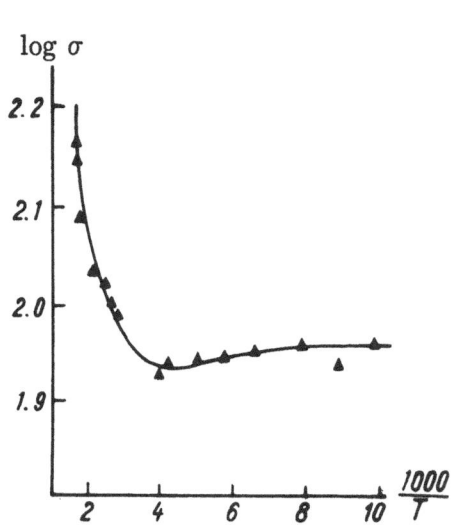

Fig. 32. Temperature dependence of the electrical conductivity of Cu_2SnSe_3.

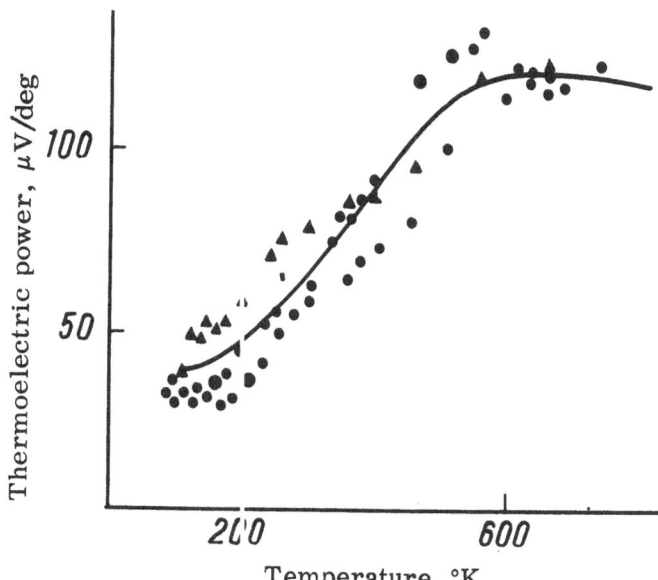

Fig. 33. Temperature dependence of the thermoelectric power of Cu_2SnSe_3.

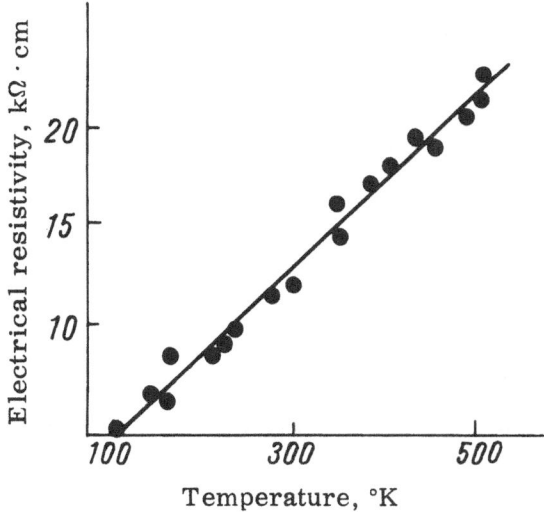

Fig. 34. Temperature dependence of the electrical conductivity of Cu_2SnTe_3.

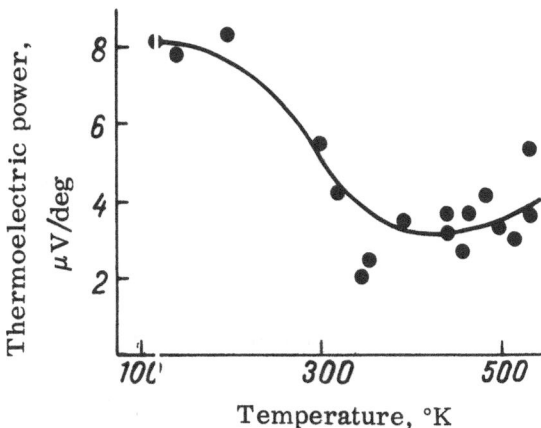

Fig. 35. Temperature dependence of the thermoelectric power of Cu_2SnTe_3.

in samples cut from the same ingots which were used to determine the thermal and elastic properties. All these samples represented the high-temperature cubic modifications because of the conditions under which they were synthesized. The results of these additional measurements are given in Table 27. These results indicate that the carrier mobility in polycrystalline samples is fairly high. Consequently, we may expect that the mobilities in single-crystal samples may be sufficiently high to satisfy the requirements of modern solid-state electronics.

Investigations have been made also of the electrical conductivity, thermoelectric power, Hall effect, and thermal conductivity of some $A_2^IB^{IV}C_3^{VI}$ compounds [162]. Figures 32 and 33 show the temperature dependences of the electrical conductivity and thermoelectric power of Cu_2SnSe_3. It has been established that Cu_2SiTe_3, Cu_2GeTe_3, and Cu_2SnTe_3 exhibit metallic conduction (Figs.

TABLE 28. Molecular Weights M, Average Atomic
Numbers Z, and Thermal Conductivities $\varkappa$
of Some $A_2^I B^{IV} C_3^{VI}$ Compounds

Compound	Z	M	$\varkappa$			
			[162]		[163]	
			$mW \cdot cm^{-1}$ $\cdot deg^{-1}$	$mcal \cdot cm^{-1}$ $\cdot deg^{-1} \cdot sec^{-1}$	$mW \cdot cm^{-1}$ $\cdot deg^{-1}$	$mcal \cdot cm^{-1}$ $\cdot deg^{-1} \cdot sec^{-1}$
Cu_2SiS_3*	20	251.40	23	5.5	—	—
Cu_2GeS_3*	23	295.88	12	2.9	7.7	1.8
Cu_2GeSe_3*	32	436.56	—	—	6.91	1.6
Cu_2GeSe_3**	—	—.	24	5.6	—	—
Cu_2GeTe_3	41	582.51	130	31.1	—	—
Cu_2SnS_3	26	341.98	28	6.7	7.33	1.7
Cu_2SnSe_3	35	482.66	35	8.4	6.60	1.6
Cu_2SnTe_3	44	628.61	144	34.5	—	—

Note. The thermal conductivity was measured in [162] using samples with an
excess of the group VI element; stoichiometric samples were used in [163].

*High-temperature modification.

**Low-temperature modification.

TABLE 29. Melting Points t_{mp}, Densities ρ, Electrical
Conductivities σ, Carrier Densities n, Carrier Mobilities μ,
and Forbidden Band Widths of ΔE of Some
Silver-Containing $A_2^I B^{IV} C_3^{VI}$
Compounds [100, 163]

Compound	t_{mp}, °C	ρ, g/cm^3	σ, $\Omega^{-1} \cdot cm^{-1}$	$n \cdot 10^{-17}$ cm^{-3}	μ, $cm^2 \cdot V^{-1}$ $\cdot sec^{-1}$	ΔE, eV	
						electrical	optical
Ag_2GeSe_3	540	4.66	27	2	850	0.9	0.91
Ag_2GeTe_3	330	5.12	92	8	720	0.25	—
Ag_2SnSe_3	490	4.86	146	10	910	0.7	0.81
Ag_2SnTe_3	315	5.76	48	5	600	0.08	—

34 and 35). Stoichiometric samples, as well as samples containing an excess of the element of
group VI, were used in these measurements.

Measurements of the temperature dependence of the thermal conductivity have indicated
that it decreases when the temperature is increased. This indicates that heat is transported
in these compounds by lattice vibrations rather than by free charge carriers. Some of the results of the measurements of the thermal conductivity, reported in [162, 163], are given in
Table 28. These results are not in agreement, possibly because of the slightly different compositions of the samples. Moreover, the extremely high values of the thermal conductivity of
tellurides give rise to some doubts about their accuracy. The lattice thermal conductivity of
the tellurides should be relatively low and heat transport by the diffusion of free carriers cannot account for such high thermal conductivities because the electrical conductivities of these
compounds are, according to [162], 3900 $\Omega^{-1} \cdot cm^{-1}$ for Cu_2GeTe_3 and 13,000 $\Omega^{-1} \cdot cm^{-1}$ for
Cu_2SnTe_3.

Ioffe's calculations [164] show that the thermal conductivity due to free carriers repre-
sents approximately 4.2 $mW \cdot cm^{-1} \cdot deg^{-1}$ for each $10^3 \Omega^{-1} \cdot cm^{-1}$. The cited values of the electrical conductivity do not agree with the reported high values of the thermal conductivity of

these tellurides. Moreover, it is worth noting that the thermal conductivity values reported in [162] are not correlated with the molecular weights (Table 23) or with the melting of these compounds.

As already reported, $A_2^I B^{IV} C_3^{VI}$ compounds containing silver have complex structures which have not yet been identified. This prevents us from making any generalizations about the properties of these compounds or their potential practical applications. Some properties of the silver compounds, taken from [100, 163] are given in Table 29.

The results of measurements of the thermoelectric properties of some $A_2^I B^{IV} C_3^{VI}$ compounds are reported in [165]. Values of the thermoelectric power, electrical conductivity, and thermal conductivity of these compounds can be used to estimate the thermoelectric figure of merit, which governs the potential applications of these materials in thermoelectric converters. It is found that the figures of merit of these compounds are poorer than those of the widely used bismuth−antimony−tellurium and bismuth−selenium−tellurium alloys.

A brief comparative review of the properties of $A_2^I B^{IV} C_3^{VI}$ compounds is given in [165].

The problem of the preparation of single crystals of $A_2^I B^{IV} C_3^{VI}$ compounds has not yet been solved. In 1963, R. V. Bakradze succeeded in growing small single crystals of Cu_2SnSe_3 by the Bridgman−Stockbarger method. Kharakhorin and Petrov [157] have measured the photo-electric properties of this compound. The preparation of single crystals of alloys containing a ternary compound of the $A_2^I B^{IV} C_3^{VI}$ type was reported by Goryunova et al. [78], who prepared single crystals of an alloy consisting of 20% Cu_2GeSe_3 and 80% 3(GaAs) by the method of gas-transport reactions, using iodine as the transport agent. This is the first known case of the preparation of single crystals of such complex composition.

The reported electrical and thermoelectric properties suggest that some $A_2^I B^{IV} C_3^{VI}$ compounds may be of practical interest. Solid solutions based on these ternary compounds may have properties which would differ in a desired manner from the parameters of the original compounds. Therefore, it is not surprising that there is much interest in solid solutions based on ternary compounds discussed in the present section.

Another reason for the interest in homogeneity (solid-solution) regions is the problem of the interaction of a ternary compound with its own components and with binary compounds in the Gibbs triangle, which contains a given ternary compound. The solution of this problem may yield information on the nature of the formation of ternary compounds, i.e., the information which would be essential in the development of techniques for preparing particular compounds.

An investigation of alloys of the quasibinary system $Cu_2GeSe_3 - Cu_2SnSe_3$, reported in [166], was carried out with these problems in mind. The first of these two compounds has the chalco-pyrite structure and the other is of the sphalerite type. The existence of chalcopyrite solid solutions has been established in the range of compositions from pure Cu_2GeSe_3 to $2Cu_2GeSe_3 \cdot Cu_2SnSe_3$. Also, solid solutions with the sphalerite structure have been found in the range from $2Cu_2GeSe_3 \cdot Cu_2SnSe_3$ to Cu_2SnSe_3 [166]. Some of the results obtained are presented in Table 30.

Averkieva and Vaipolin [149] showed that the system $Cu_2GeSe_3 - Cu_2GeTe_3$ forms non-equilibrium substitutional solid solutions throughout the whole range of concentrations. Unfortunately, the electrical, optical, and thermoelectric properties of these solid solutions have not yet been investigated.

The solubility of ternary and binary diamond-like semiconductors has been investigated by several workers [149, 152]. It is shown in [149] that Cu_2GeSe_3 can form solid solutions with

TABLE 30. Some Physicochemical Properties of Quasibinary
$Cu_2GeSe_3 - Cu_2SnSe_3$ Homogeneous Alloys [166]

Alloy composition	Structure	Lattice parameters, nm (Å)		Melting point (crystalliza- tion range), °C
		a	c	
Cu_2GeSe_3	Chalcopyrite	0.558 (5.58)	1.096 (10.96)	788
$3Cu_2GeSe_3 \cdot 1Cu_2SnSe_3$	do.	0.560 (5.60)	1.108 (11.08)	704—765
$2Cu_2GeSe_3 \cdot 1Cu_2SnSe_3$	Sphalerite	0.560 (5.60)	—	701—769
$1Cu_2GeSe_3 \cdot 1Cu_2SnSe_3$	do.	0.561 (5.61)	—	687—741
$1Cu_2GeSe_3 \cdot 2Cu_2SnSe_3$	do.	0.563 (5.63)	—	695—744
$1Cu_2GeSe_3 \cdot 3Cu_2SnSe_3$	do.	0.565 (5.65)	—	704—745
Cu_2SnSe_3	do.	0.570 (5.70)	—	709

gallium arsenide, and that these solutions have the sphalerite structure. Goryunova et al. [152] were able to prepare an alloy of 25% Ag_2SnTe_3 + 75% InSb with the zinc-blende structure and they have found, by microstructure analysis, that the concentration of a second phase in this alloy does not exceed 2%.

Reports of investigations of the interaction of ternary compounds with components of these compounds and with binary compounds of similar composition can be found in [22, 149, 158]. It is pointed out in [149] that Cu_2GeSe_3 has a very narrow homogeneity region, which is in practice limited to the stoichiometric composition. However, solid solutions have been found in the $Cu_2GeSe_3 - Ge$ system and the range of existence of solid solutions is fairly wide: from Cu_2GeSe_3 to $Cu_2Ge_2Se_3$, i.e., up to 14% of Ge can be dissolved in the ternary compound [153]. In view of this, investigations of ternary semiconducting compounds should be carried out under conditions permitting the formation of variable-composition phases because they may form interesting solid solutions.

In [158] there is a description of the quasibinary system $Cu_2Se - GeSe_2$, which satisfies the condition of normal valence. It has been found that the tetragonal symmetry exists throughout the whole range of concentrations. The homogeneity region of this system lies near the compound Cu_2GeSe_3 and is extremely narrow. Many alloys of this system exhibit photosensitivity in the thin-film form, which may be of practical importance.

Luzhnaya [167] pointed out that studies of quasibinary subsystems are insufficient to obtain information on ternary systems, particularly when the components are soluble in the ternary compound. It is necessary to investigate whole regions in the Gibbs triangle. However, such investigations have not yet been undertaken for the systems described in the present monograph. The absence of such information is an obstacle in the development of techniques for the preparation of perfect single crystals of ternary compounds, because current attempts to prepare such single crystals are usually based on the experience accumulated in studies of compounds such as those of the $A^{III}B^V$ type. Even the very few physicochemical data available at present show that the nature of the formation of ternary and binary compounds may differ considerably. Therefore, it is likely that in order to prepare perfect single crystals of ternary compounds it would be necessary to develop special techniques.

Another problem relating to $A_2^I B^{IV} C_3^{VI}$ compounds is their interaction with impurities and the influence of impurities on the electrical properties; systematic studies of the physicochemical, electrical, optical, and other properties of doped and undoped crystals are required because the published data are contradictory (this applies particularly to the electrical properties), because of the considerable dependence of the properties of these compounds on the methods used to synthesize them.

$A_3^I B^V C_4^{VI}$ and $A^I B_2^{IV} C_3^V$ Compounds

Ternary compounds of elements of the first, fifth, and sixth group of the periodic table are of special interest because the cation belongs to the fifth group.

Naturally occurring minerals of this group are known: they include enargite (Cu_3AsS_4) and famatinite, which is a phase based on Cu_3SbS_4 with antimony partly replaced with arsenic and copper partly replaced with iron or zinc. The structures of these minerals were first described by de Jong [168], who established their isomorphism and similarity of the lattice period, as well as by Hermann, Lohrmann, and Philipp [169]. Investigations of the structure of enargite and famatinite have been reported also in [170-172]. Other workers [183, 184] have established that enargite has the orthorhombic structure with the unit lattice parameters: $a = 0.646$ nm (6.46 Å), $b = 0.743$ nm (7.43 Å), $c = 0.618$ nm (6.18 Å), V = 0.2965 nm^3 (296.5 Å^3), Z = 2 (space group Pnm).

A mineral of the composition Cu_3PS_4 has also been found to have the orthorhombic structure with periods practically identical with the lattice periods of enargite and with the same space group [185].

It is also interesting to note the natural occurrence as a mineral of Cu_3VS_4, because its composition includes a transition metal. According to [168], Cu_3VS_4 has a cubic lattice with the following unit cell parameters: $a = 1.075$ nm (10.75 Å), V = 1.24 nm^3 (1240 Å^3), Z = 8, and the space group Fm^3m, formed by VS_4 tetrahedra. However, in another report [185] this compound is assigned the lattice parameter $a = 5.37$ Å and the space group $P43m$.

The structure of enargite (Cu_3AsS_4) is represented in [186] in the form of layers consisting of tetrahedra formed by copper, arsenic, and sulfur atoms.

Cu_3AsS_4 and Cu_3SbS_4 were synthesized first by Wernick and Benson [173], simultaneously with a large group of compounds of the $A_3^I B^V C_4^{VI}$ and $A^I B^V C_2^{VI}$ types. These compounds were synthesized by the fusion of the elements with an excess (1%) of arsenic and sulfur in a quartz ampoule filled with nitrogen at a pressure of two-thirds of the atmospheric pressure; the temperature was raised very slowly.

According to the results reported by Wernick and Benson [173], the melting points of Cu_3AsS_4 and Cu_3SbS_4 are, respectively, 655 and 555°C and these substances are semiconductors with a forbidden band not wider than 1 eV. Moreover, they [173] synthesized Cu_3SbSe_4, whose melting point is 425°C. According to Wernick and Benson, the low vapor pressures of Cu_3AsS_4, Cu_3SbS_4, and Cu_3SbSe_4 at temperatures exceeding their melting points by 100 deg C suggest that it should be easy to purify them by zone recrystallization and to prepare single crystals. However, all our attempts to grow single crystals of these compounds were unsuccessful. Cu_3AsS_4 has the orthorhombic structure, similar to that of wurtzite, while Cu_3SbS_4 and Cu_3SbSe_4 have the same cubic structure [173].

A US patent, No. 2,882,192 of 1959, describes the synthesis of Cu_3AsS_4 and Cu_3SbS_4 in a three-zone vertical furnace, using a quartz ampoule containing a covered crucible in which the elements are placed. The ampoule is filled with nitrogen to a pressure of 202-303 kN/m^2 (2-3 atm). It has been reported that this method produces polycrystalline ingots with grain dimensions ranging from 2.5 to 12.5 mm. Samples of these compounds have the n-type conduction. Doping with chlorine, iodine, and manganese alters the sign of conduction. The forbidden band widths of these compounds (about 0.8 eV) suggest that they might find applications in infrared technology and in the manufacture of point-contact or planar diodes. The patent itself describes point-contact diodes prepared from crystals of these two compounds obtained by the described synthesis technique.

TABLE 31. Magnetic Susceptibilities
of Some $A_3^I B^V C_4^{VI}$ Compounds

Compound	Magnetic susceptibility		
	Specific χ, m³/g	Molar χ_m, m³/mole	Mean atomic χ_a, m³/g-atom
Cu_3AsS_4	0.28	110	13.7
Cu_3SbS_4	0.12	53	6.6
Cu_3AsSe_4	0.15	87	10.9
Cu_3SbSe_4	0.20	130	16.3

TABLE 32. Physicochemical, Thermal, and Elastic
Properties of Some $A_3^I B^V C_4^{VI}$ Compounds

Property	Cu_3AsS_4	Cu_3SbS_4	Cu_3AsSe_4	Cu_3SbSe_4
Average atomic number	23	25.25	32	34.25
Molecular weight	393.794	440.644	581.37	628.22
Structure	Orthorhomb.	Sphalerite	Sphalerite	Sphalerite
Unit cell parameters, nm (Å):				
a	0.646 (6.46)	5.28	5.5	5.6
b	0.743 (7.43)			
c	0.618 (6.18)			
Z	2			
Boiling point, °C	655	555	460	448
Density, g/cm³:				
theoretical	4.34	4.90	5.74	5.94
experimental	4.37	—	5.61	6.0
Thermal conductivity mcal·cm⁻¹·sec⁻¹·deg⁻¹ (mW·cm⁻¹·deg⁻¹) . . .	7.22 (30.2)	—	4.55 (19)	3.5 (14.6)
Linear expansion coeff. ×10⁶, deg⁻¹	3.2	—	9.5	12.4
Debye temperature, °K.	—	—	169	131

TABLE 33. Electrical Properties of Some
$A_3^I B^V C_4^{VI}$ Compounds

Property	Cu_3AsS_4	Cu_3AsSe_4	Cu_3SbSe_4
Electrical conductivity, $\Omega^{-1}·cm^{-1}$.	0.095	215	50
Hall coefficient, cm³/C	0.093	2.77	3.10
Carrier density, cm⁻³	$7.8·10^{19}$	$2.7·10^{18}$	$2.4·10^{18}$
Carrier mobility, cm²·V⁻¹·sec⁻¹ . . .	0.008	505	135
Thermoelectric power, u V/deg	130	120	200
Thermoelectric figure of merit ×10³, deg⁻¹	—	0.58	0.13
Forbidden band width, eV:			
photoelectric, at 77°K	1.48	0.65	—
optical, at 300°K	1·24	0.88	0.31
from elec. meas.	0.8	0.76	0.42

Busch and Hulliger [174] considered the principles of the formation of the tetrahedral and octahedral coordination in compounds and gave values of the lattice parameters of Cu_3AsSe_4 and Cu_3SbSe_4 without any indication how they prepared these compounds. According to their data, these compounds have tetragonal structures similar to that of sphalerite (and evidently close to the chalcopyrite lattice) with the following parameters:

Cu_3AsSe_4

$$a = 0.557 \pm 0.0003 \text{ nm } (5.570 \pm 0.003\text{Å}),$$
$$c = 1.0957 \pm 0.0005 \text{ nm } (10.957 \pm 0.005\text{Å}),$$
$$c/a = 1.967 \pm 0.002;$$

Cu_3SbSe_4

$$a = 0.5654 \pm 0.0003 \text{ nm } (5.654 \pm 0.003\text{Å}),$$
$$c = 1.1256 \pm 0.0005 \text{ nm } (11.256 \pm 0.005\text{Å}),$$
$$c/a = 1.991 \pm 0.002.$$

It is worth considering the compound Cu_3VSe_4, whose structure is similar to the mineral sulvanite with the lattice parameter $a = 0.5569 \pm 0.0002$ nm (5.569 ± 0.002 Å) [174].

Alieva and Pinsker [187] carried out an electron-diffraction study of the structure of Cu_3SbS_4 films, prepared by evaporation on freshly cleaved rocksalt substrates. They reported the existence of ordered and disordered structures. The lattice constant of the ordered phase is $a = 1.074$ nm (10.74 Å) and its space group is $Fm3m$; the disordered phase has the zinc-blende structure with $a = 0.528$ nm (5.28 Å) and $Z = 1$.

There are several possible ordered configurations of atoms in the crystal lattices of $A_3^I B^V C_4^{VI}$ compounds, as pointed out by Newman [175]. By way of example, we can point out that Goodman [10] as well as Goryunova et al. [176] attribute a diamond-like structure to Cu_3AsSe_4.

The optical and photoelectric properties of $A_3^I B^V C_4^{VI}$ compounds are described in [188]. Values of the electrical properties, thermal conductivity, thermal expansion, and Young's modulus are given in [47, 178].

The results of measurements of the magnetic susceptibility of Cu_3AsS_4, Cu_3SbS_4, Cu_3AsSe_4, and Cu_3SbSe_4 can be regarded as an indirect confirmation that the crystal structure of Cu_3AsS_4 differs from the structure of the other three compounds, as shown in Table 31, which is taken from R. Annamamedov's dissertation for candidate's degree (Moscow Steel and Alloys Institute, 1967).

Samples of compounds of this type usually have p-type conduction. When they are doped with zinc, the sign of conduction is reversed. The temperature dependences of the electrical conductivity and Hall effect of undoped and doped samples of these compounds were determined by Annamamedov et al. [179].

A study of the solubility in the Cu_3AsSe_4–InAs system is reported in [152]. It has been found that an alloy consisting of 25% Cu_3AsSe_4 and 75% InAs contains not more than 5% of a second phase. This suggests the possibility of obtaining solid solutions in this system, which is of definite practical interest.

Various properties of $A_3^I B^V C_4^{VI}$ compounds, taken from the results published in [100, 174, 177-179, 188, 189], are given in Tables 32 and 33.

It is worth mentioning another group of ternary compounds of elements of the first, fifth, and sixth groups: they are the compounds of the $A^IB^VC_2^{VI}$ type. In accordance with the rules for the formation of tetrahedral phases most of these compounds have the rocksalt-type structure, i.e., a detailed analysis of these compounds is outside the scope of the present monograph, but we shall consider them briefly.

There have been reports of the synthesis of $AgAsSe_2$ and $AgAsTe_2$ [180] and of alloys of the system $AgSbSe_2-AgSbTe_2-AgBiSe_2-AgBiTe_2$ [181]. An investigation of alloys of the $AgBiTe_2-PbTe$ system is reported in [190]. The thermoelectric, thermal, and galvanomagnetic properties of $AgSbTe_2$ are reported in [182], while the method for synthesis of single crystals of this compound is described in [191]. The thermoelectric properties and the figures of merit of the solid solutions $AgSbSe_2-AgSbTe_2$, $AgBiSe_2-AgBiTe_2$, $CuSbSe_2-CuSbTe_2$, and $CuBiSe_2-CuBiTe_2$ can be found in [124].

The interest in these compounds and their solid solutions is due to their relatively low thermal conductivity and high values of the thermoelectric power and electrical conductivity. The carrier mobility in a sample of $AgSbSe_2$, purified by zone melting, reaches 1500 $cm^2 \cdot V^{-1} \cdot sec^{-1}$ [180].

We shall now consider compounds of the $A^IB_2^{IV}C_3^V$ type; we find that extremely few of these ternary diamond-like phases are known at present.

Goryunova and Sokolova [192] have described a method for synthesis of $CuGe_2P_3$. The synthesis consisted of the fusion of the elements in an evacuated and sealed quartz ampoule. It is necessary to use vibrational stirring in order to prevent the explosion of the ampoule because of the high vapor pressure of phosphorus above the melt. The x-ray diffraction data show that $CuGe_2P_3$ has the sphalerite structure with a lattice constant $a = 0.532$ nm (5.32 Å); its microhardness is 8338.5 ± 196.2 MN/m^2 (850 ± 20 kgf/mm^2) and the samples are gray with a metallic luster. No solubility has been found in the $CuGe_2P_3-InP$ system.

Folberth and Pfister [127] reported that $CuGe_2P_3$ and $CuSi_2P_3$ have the sphalerite structure with a disordered distribution of atoms in the cation sublattice. The lattice parameters are 0.525 nm (5.25 Å) for $CuSi_2P_3$ and 5.37_5 Å for $CuGe_2P_3$.

No $A_3^IB^VC_4^{VI}$ compounds, containing arsenic or antimony as the element of the fifth group, are known. Therefore, there is interest in trying to synthesize homogeneous alloys of the composition $A^IB_2^{IV}(As, Sb)_3-A^{III}B^V$. This was attempted by Goryunova, Voitsekhovskii, and Prochukhan [152], who were able to prepare an alloy with 50% $CuGe_2As_3$ and 50% InAs with the sphalerite structure, containing less than 15% of a second phase, as well as a single-phase alloy consisting of 25% $AgSn_2Sb_3$ and 75% InSb with the same structure.

Goryunova and Chiang Ping-Hsi [153] considered the problem of the solubility of germanium in $CuGe_2P_3$. They observed the formation of solid solutions of $CuGe_xP_3$ composition, where $x = 2-5$. Their conclusion was that the dissolved atoms of "amphoteric" germanium can occupy both cation and anion sites in accordance with the charge compensation condition.

Sokolova [193] investigated some alloys of the copper-germanium-phosphorus system along the $Cu_3P-Ge_3P_4$ and CuP_3-Ge sections, the former of which corresponds to the condition of normal valence and the latter to the four-electron condition. Sokolova has found a homogeneity region within the limits of which the lattice parameter of the alloys ranges from 0.538 to 0.548 nm (from 5.38 to 5.48 Å). She reported that the lattice constant of $CuGe_2P_3$, which has the sphalerite structure, is 0.538 nm (5.38 Å) and its microhardness is in agreement with the values reported in [192]. Sokolova [193] pointed out that ternary $A^IB_2^{IV}C_3^V$ compounds may have a homogeneity region but this region is not as wide as in quaternary systems.

Sokolova and Tsvetkova [194] considered the synthesis of $A^I B_2^{IV} C_3^V$ compounds with copper, silver, or gold as the element of the first group and silicon, germanium, or tin as the elements of the fourth group. They have obtained $CuSi_2P_3$ with the sphalerite structure and lattice parameters equal to those reported in [127]. Attempts to synthesize compounds of gold have not been successful. The main phase in $CuSn_2P_3$ has the sphalerite structure but that in $AgSn_2P_3$ has a complex structure which has not yet been interpreted. An alloy corresponding to the composition $CuGe_2P_3$ has been found to contain 3-5% of a second phase. The main phase in this alloy has the sphalerite structure with a lattice constant 0.53_8 nm (5.3_8 kX). The melting point of this alloy is 800°C. Moreover, thermal analysis shows an additional thermal effect at 759°C.

$AgGe_2P_3$ has the bcc lattice and its microhardness is 7161.3 ± 196.2 MN/m^2 (730 ± 20 kgf/mm^2) [194]. According to thermal analysis, there is only one transition at 742°C.

The results, reported in [194], of a physicochemical investigation of alloys of the $CuGe_2P_3$−Ge system confirm the existence of solid solutions with the sphalerite structure and a lattice constant of the homogeneous alloys varying with the concentration in accordance with the Vegard law [153]. The $CuGe_2P_3$−Ge_3P_4 system also has a wide range of existence of homogeneous phases with the sphalerite structure.

It is worth noting the considerable discrepancy between the lattice parameters of alloys of the $CuGe_2P_3$ composition obtained by different workers. It is possible that $CuGe_2P_3$ is a variable-composition phase.

The electrical, thermoelectric, photoelectric, and thermal properties of $A^I B_2^{IV} C_3^V$ compounds and of solid solutions based on these compounds have not yet been investigated and therefore no predictions can be made about possible practical applications of these substances.

$A^{II} B^{IV} C_2^V$ Compounds

Ternary compounds of this type can be regarded as ternary analogs of $A^{III} B^V$ binary compounds, obtained from the latter by the cross substitution method [196]. In 1956, Austin, Goodman, and Pengelly [89] suggested that, if these compounds exist, they should have the chalcopyrite structure. The same conclusion was reached by Goryunova [195].

In 1957, Goodman [197] reported the synthesis of seven compounds of $A^{II} B^{IV} C_2^V$ type having the chalcopyrite structure. The lattice parameters of some compounds of this type were reported by Folberth and Pfister [196] (Table 34). The crystal structures of some of these compounds were also discussed by Pfister [198]. Analyses have been made of the x-ray reflections in the Debye diffractions patterns of these compounds on the assumption that their space group is D_{2d}^{12} ($I\bar{4}2d$), typical of the chalcopyrite structure. The following types of indices were found for $A^{II} B^{IV} C_2^V$ compounds [271]:

$$(hkl) \text{ only when } h+k+l=2n,$$
$$(0kl) \text{ only when } k+l=2n,$$
$$(hkl) \text{ only when } 2h+l=4n \text{ and } l=2n,$$
$$(h00) \text{ only when } h=2n.$$

Some of the results reported in [271] are given in Table 34.

By analogy with the $A^{III} B^V$ compounds, we could expect some $A^{II} B^{IV} C_2^V$ compounds to have the wurtzite structure. Rabenau and Eckerlin [33] have shown that $BeSiN_2$ has the wurtzite structure. According to these authors [33], $BeSiN_2$ has the following lattice parameters: $a = 0.287_2 \pm 0.0004$ nm ($2.87_2 \pm 0.004$ Å), $c = 0.467_4 \pm 0.0004$ nm ($4.67_4 \pm 0.004$ Å), $c/a = 1.62_7$. The fol-

TABLE 34. Lattice Parameters and Unit Cell Volumes
of Some $A^{II}B^{IV}C_2^V$ Compounds [196, 198]

Compound	Lattice parameters		Unit cell volume nm³ (Å³)
	a, nm (Å)	c/a	
$ZnGeP_2$	0.546 ± 0.001 (5.46 ± 0.01)	1.97 ± 0.01	32 (320.5)
$ZnSiP_2$	0.539_8 (5.39)	1.93_4	—
$ZnGeAs_2$	0.567 ± 0.0002 (5.670 ± 0.002)	1.967 ± 0.002	35.86 (358.6)
$CdGeAs_2$	0.5942 ± 0.0002 (5.942 ± 0.002)	1.889 ± 0.002	—
$CdSnAs_2$	0.6092 ± 0.0002 (6.092 ± 0.002)	1.957 ± 0.002	44.24 (442.4)

lowing values of density are reported: calculated, 3.24 g/cm³, pycnometric, 3.12 g/cm³. Rabenau and Eckerlin [33] conclude that $BeSiN_2$ can be regarded as an analog of aluminum, gallium, or indium nitrides.

The crystal structure of $A^{II}B^{IV}C_2^V$ compounds depends considerably on the method used in synthesis and subsequent heat treatment [191, 199], as already pointed out in a discussion of $A_2^I B^{IV}C_3^{VI}$ compounds. By way of illustration, we shall consider the structure of $ZnSnAs_2$. This compound usually has the sphalerite structure [23] with the unit cell parameter $a = 0.5851$ nm (5.851 Å). To explain this observation, Folberth and Pfister [23] used the relative polarization of the valences and concluded [200] that the sphalerite structure of $ZnSnAs_2$ is due to a greater similarity of the polarizability coefficients of bonds in this compound than in other compounds of the $A^{II}B^{IV}C_2^V$ type. However, Gasson et al. [201] showed that $ZnSnAs_2$, like many other compounds of this type, has two crystallographic modifications: a low-temperature chalcopyrite structure with the parameters $a = 0.5852 \pm 0.0001$ nm (5.852 ± 0.001 Å), $c = 1.1705$ nm (11.705 Å), $c/a = 2.0000 \pm 0.0012$, and a high-temperature form with the sphalerite structure; a transition from the ordered to disordered distribution of atoms in the cation sublattice occurs at a temperature close to 650°C.

Later, Pfister [208] obtained results practically identical with those reported by Gasson et al. [201]. This problem was considered also by Folberth and Pfister in [127], particularly the ratios of the ionic $(r_a/r_b)_{ion}$ and covalent $(r_a/r_b)_{cov}$ radii of $A^{II}B^{IV}C_2^V$ compounds. Folberth and Pfister [127] concluded that, since the ratios $(r_a/r_b)_{ion}$ for these compounds are greater than the ratios $(r_a/r_b)_{cov}$, the polarization of the $A^{II}-B^V$ bonds is stronger than the polarization of the $A^{IV}-B^V$ bonds; also, when the difference between these ratios is large, the chalcopyrite structure is expected but when the difference is small, the sphalerite structure should be observed. However, as already pointed out, this conclusion is not sufficiently general because, for example, $ZnSnAs_2$, $ZnSiP_2$, and $CdSnAs_2$, all have the chalcopyrite structure although the difference between the radius ratios is small (0.10) for $ZnSnAs_2$ [201] and large (0.69) for $ZnSiP_2$ and $CdSnAs_2$.

$MgGeP_2$ was synthesized first by Folberth and Pfister [127]. They found that this compound has the cubic structure of the sphalerite type with the lattice constant of 0.565_2 nm (5.65_2 Å).

The synthesis of polycrystalline samples of $A^{II}B^{IV}C_2^V$ compounds is usually carried out (with the exception of refractory compounds) by the fusion of stoichiometric amounts of the elements in evacuated and sealed quartz ampoules. Some workers have found that samples prepared in this way have many cracks. In our opinion, the formation of cracks can be avoided by reducing the rate of cooling of the melt. Coarse-grain ingots may be prepared by the Bridgman directional (horizontal) crystallization method [21].

Refractory compounds can be synthesized by the powder metallurgy method. Thus, to obtain $BeSiN_2$, stoichiometric amounts of binary nitrides are compacted into a tablet, placed in a

TABLE 35. Melting Points t_{mp}, Lattice Parameters,
Microhardness H, and Thermal Conductivity $\varkappa$
of Some $A^{II}B^{IV}C_2^V$ Compounds

Compound	t_{mp}, °C		Lattice parameters [286]		H, MN/m² (kgf/mm²) [211]	$\varkappa$, mW·cm⁻¹·deg⁻¹ (mcal·cm⁻¹·sec⁻¹·deg⁻¹)	
	[211]	[212, 287]	a, nm (Å)	c/a		[286, 211]	[212, 287]
ZnSiAs₂	1038	—	0.559 (5.59)	1.94	8927 (910)	—	—
ZnGeP₂	1020	—	0.546 (5.46)	1.97	8614 (980)	41.8 (10)	—
ZnGeAs₂	850	875	0.567 (5.67)	1.97	6867 (700)	125 (30)	114 (27)
ZnSnAs₂	—	775		—	—	—	74 (18)
CdGeP₂	776	—	0.574 (5.74)	1.88	8338 (850)	83.6 (20)	—
CdGeAs₂	665	670	0.594 (5.94)	1.89	4444 (453)	41.8 (10)	40 (9.5)
CdSnAs₂	615	595	—	—	3875 (395)	—	70 (17)

boron nitride crucible, and heated in a stream of ammonia vapor at temperatures of 1750-1800°C [33]. The method of gas-transport reactions is promising in the synthesis of $A^{II}B^{IV}C_2^V$ compounds. Details of synthesis of particular compounds will be given as part of the description of these compounds.

The problem in the preparation of perfect single-crystal samples of $A^{II}B^{IV}C_2^V$ compounds is still far from being solved. This is primarily because of the insufficient knowlege of the phase diagrams of the $A^{II}-IV-C^V$ ternary system and the consequent absence of data on the nature of formation of compounds. In this connection, we must mention the very interesting research undertaken by Borchers and Maier [38], who investigated thirteen quasibinary sections of the zinc—tin—arsenic ternary system. According to them, ZnSnAs₂ is formed by a peritectic reaction and has a solid-state phase transition near 645°C.

In spite of the absence of data on the nature of the formation of $A^{II}B^{IV}C_2^V$ compounds, considerable success has been achieved in the preparation of single crystals of these compounds, although the selection of the methods for growing single crystals has been largely intuitive. The most interesting, in our opinion, is the method for growing single crystals developed at the Physicotechnical Institute of the USSR Academy of Sciences [55], which combines the synthesis and single-crystal growth processes. A characteristic feature of this method is that the directional (horizontal) Bridgman crystallization is used with the same temperature gradient along the whole ampoule. Evidently, this point is of cardinal importance in the crystallization process since it ensures a dynamic equilibrium between the atoms (or molecules in the solid and molten parts of the charge, because of the constancy of the thermal particle-velocity gradient on both sides of the crystallization front. The apparatus used for this purpose is shown schematically in Fig. 13. Using this apparatus, single crystals measuring 8 × 10 × 60 mm have been grown for a large number of $A^{II}B^{IV}C_2^V$ compounds. This has made it possible to carry out a wide range of electrical and optical measurements on single crystals of these compounds [76, 209, 210].

In [211-213], there is a report of a systematic investigation of the physicochemical and physical properties of a large group of $A^{II}B^{IV}C_2^V$ compounds. Table 35 lists values of the investigated properties for some of these compounds.

Let us now consider in more detail a few of these compounds. From the practical point of view, the most interesting is CdSnAs₂, which is the ternary analog of indium arsenide. The crystal lattice parameters of CdSnAs₂ are: a = 0.6084 nm (6.084 Å), c = 1.1916 nm (11.916 Å), c/a = 1.957 [202]. These values are somewhat lower than those given in [198], with the exception of the ratio c/a, which has the same value in both cases. Other investigations of CdSnAs₂ [201, 206] have shown that it has the chalcopyrite structure and the following lattice parameters: $a = 0.6093 \pm 0.0001$ nm (6.093 ± 0.001 Å) and $c/a = 1.957 \pm 0.003$. Large single crystals of this compound have been prepared [206] and their microhardness and electrical properties measured. It has been found that $CdSnAs_2$ has the highest molecular weight among known compounds of this type. Comparison of the values of the forbidden band width and microhardness of $CdSnAs_2$ [0.26 eV and 3875 MN/m^2 (395 kgf/mm^2)] and InAs [0.45 eV and 3237 MN/m^2 (330 kgf/mm^2)] indicates that the covalence of the binding forces in $CdSnAs_2$ is stronger than in indium arsenide [206].

Rosenberg and Strauss [202] investigated the electrical properties and crystal structure of $CdSnAs_2$. They observed a very high carrier mobility, reaching 5600 $cm^2 \cdot V^{-1} \cdot sec^{-1}$ (at 300°K), in polycrystalline samples with an electron density of $2.7 \cdot 10^{18}$ cm^{-3}; their investigations indicated that even higher values of the mobility could be reached. This is suggested primarily by the constancy of the Hall coefficient in the 78-300°K range and the small value of the thermoelectric power (54 μV/deg at 300°K).

The high values of the mobility in $CdSnAs_2$ have stimulated further detailed investigations of this compound. Strauss and Rosenberg [21] have used various methods of synthesis of $CdSnAs_2$. The synthesis has been carried out in quartz ampoules using stoichiometric amounts of the components. Strauss and Rosenberg have employed directional Bridgman crystallization, slow cooling, and quenching. They have found that when the melt is cooled at a rate of about 1 deg/min, coarse-grain ingots are obtained in which individual crystallites are up to 1 cm size. Their measurements of the electrical conductivity and of the Hall effect have indicated a mobility of 12,000 $cm^2 \cdot V^{-1} \cdot sec^{-1}$ for an electron density of $5.5 \cdot 10^{17}$ cm^{-3}. It has also been found that an increase in the carrier density shifts the optical absorption edge in the direction of the shorter wavelengths, as observed earlier for n-type InSb [203] and n-type InAs [214]. The results of optical measurements indicate that the forbidden band width of $CdSnAs_2$ at room temperature is 0.23 eV. According to Strauss and Rosenberg [21], comparison of the values of the carrier density and mobility with the value of the thermoelectric power of several binary and ternary diamond-like compounds confirms the analogy of the energy band structure of $CdSnAs_2$ and some binary compounds, from which they conclude that the conduction band of $CdSnAs_2$ may be nonparabolic, as found earlier by Kane [204] for indium antimonide and by Harman [215] for mercury selenide. More details about the band structure of $A^{II}B^{IV}C_2^V$ compounds will be given later.

The results of measurements of the electrical conductivity and the Hall effect of $CdSnAs_2$, carried out in the temperature range 77-840°K, on polycrystalline and single-crystal samples of $CdSnAs_2$ are reported also in [216]. It has been found that, at an electron density of 10^{17} cm^{-3}, their mobility reaches 22,000 $cm^2 \cdot V^{-1} \cdot sec^{-1}$. It must be mentioned that an increase of the carrier density to $1.1 \cdot 10^{18}$ cm^{-3} is accompanied, according to the results reported by Gasson et al. [201], by a fall of the mobility to 3140 $cm^2 \cdot V^{-1} \cdot sec^{-1}$.

The thermal conductivity of $CdSnAs_2$ was first measured by Gasson et al. [201]. Their values of the thermal conductivity of $CdSnAs_2$ differ by a factor of 2 from those reported in [100]. Since the thermal conductivity is a phenomenological property, it is very likely that this difference between the results is due to the characteristic features of synthesis employed by Gasson et al. [201], who melted stoichiometric amounts of the components in graphite crucibles, placed in quartz ampoules, and followed synthesis by annealing and slow cooling.

The magnetic susceptibility of $CdSnAs_2$ was investigated by Matyáš and Höschl [207]. They discovered that this compound, like many other diamond-like materials, is diamagnetic and its molar magnetic susceptibility is -112 m^3/mole. The electrical measurements of Matyáš and Höschl have confirmed the value of the forbidden band width of $CdSnAs_2$ obtained by Goryunova, Mamaev, and Prochukhan [206]. The effective electron mass found by Matyáš and Höschl is $m^* = 0.02$ m_0.

Goryunova et al. [209, 210] investigated single crystals of $ZnSnAs_2$, having p-type conduction. They established that this compound has the chalcopyrite structure with the parameters: $a = 0.58515 \pm 0.00005$ nm (5.8515 ± 0.0005 Å), $c = 1.1703 \pm 0.0001$ nm (11.703 ± 0.001 Å). Measurements of the temperature dependence of the electrical conductivity and of the Hall effect have shown that $ZnSnAs_2$ samples exhibit impurity (extrinsic) conduction up to about 500°K, while intrinsic conduction is observed above 800°K. Single crystals of this compound are transparent in the wavelength range 1.5-3 μ. The optical value of the forbidden band width is 0.65 eV, which is in agreement with the results obtained by Gasson et al. [201]. The mobility of holes is approximately 90 $cm^2 \cdot V^{-1} \cdot sec^{-1}$ in the impurity conduction region for a hole density of about $2 \cdot 10^{18}$ cm^{-3}.

All the temperature dependences of the Hall coefficient of $ZnSnAs_2$ samples exhibit a maximum near 150-200°K. This maximum may be due to the presence of an additional energy band near the top of the valence band of $ZnSnAs_2$, with an effective density-of-states mass in this additional band less than that in the main valence band. A similar effect has been observed at low temperatures for gallium arsenide crystals [219]. An increase in the Hall coefficient near the transition from the impurity to the intrinsic conduction region can be attributed to carrier transitions from one band to the other [220].

Kesamanly, Nasledov, and Rud' [210] have determined the temperature dependence of the thermoelectric power and of the transverse Nernst−Ettingshausen effect using apparatus described by Emel'yanenko and Trishin [223]; the results have been used to determine the effective mass and the nature of carrier scattering in $ZnSnAs_2$ crystals. These results have indicated that the power exponent in the dependence of the mean free path of carriers on their energy decreases from about 2 at 100°K (scattering of carriers by ionized impurity atoms) to 1 at 300°K (scattering by polar lattice vibrations). Values of the reduced Fermi level indicate that the effective carrier mass in $ZnSnAs_2$ is $m^* = 0.13 m_0$.

The thermal conductivity of $ZnSnAs_2$ was first measured by Gasson et al. [201]. The results of their measurements are in good agreement with those reported in [100] ($6 \cdot 10^{-2}$ and $5.8 \cdot 10^{-2}$ $W \cdot cm^{-1} \cdot deg^{-1}$, respectively). Gasson et al. [201] checked whether $ZnSnAs_2$ could be used in thermoelectric applications because an investigation of the electrical properties of polycrystalline samples has indicated a low carrier mobility (25 $cm^2 \cdot V^{-1} \cdot sec^{-1}$), which would mean that this material would not be very promising in electronics applications.

A calculation of the thermoelectric figure of merit, using a formula given by Chasmar and Stratton [205]:

$$\beta = 2 \left(\frac{2\pi m^* kT}{h^2} \right)^{3/2} \left(\frac{k}{e} \right)^2 \frac{e \cdot T \cdot \mu}{\varkappa}$$

(where m^* is the effective carrier mass; k is the Boltzmann constant; h is the Planck constant; e is the electron charge; T is the absolute temperature; μ is the carrier mobility; $\varkappa$ is the thermal conductivity) has shown that the value of this figure for $ZnSnAs_2$ is approximately two orders of magnitude smaller than that for a compound such as Bi_2Te_3, for which $\beta \approx 0.25$.

According to the results of optical measurements, the forbidden band width of $ZnSnAs_2$ is 0.6 eV and the effective carrier mass, calculated from the thermoelectric power, is $0.5 m_0$ [201].

Single crystals of $CdGeAs_2$ were prepared first by Goryunova and her colleagues [225]. According to their results, this compound has the chalcopyrite structure with the parameters: $a = 0.59427 \pm 0.00005$ nm (5.9427 ± 0.0005 Å), $c = 1.12172 \pm 0.00005$ nm (11.2172 ± 0.0005 Å), and $c/a = 1.8875 \pm 0.0005$. The melting point of $CdGeAs_2$ is 665°C and this microhardness is 4620.5 ± 98 MN/m^2 (471 ± 10 kgf/mm^2). According to optical measurements, the room-temperature forbidden band width is 0.53 eV.

The thermoelectric power of $CdGeAs_2$ has been measured using a device described in [220]. At room temperature, the thermoelectric power is 190 $\mu V/deg$. The electron mobility in a single crystal of $CdGeAs_2$ is about 900 $cm^2 \cdot V^{-1} \cdot sec^{-1}$ for an electron density of 10^{17} cm^{-3} and the hole mobility is about 25 $cm^2 \cdot V^{-1} \cdot sec^{-1}$. The relatively low values of the thermoelectric power and of the carrier density in this semiconductor indicate electron gas degeneracy, which would suggest a small effective mass. In view of this, we may expect a considerable improvement in the carrier mobility of $CdGeAs_2$ samples with fewer defects.

Leroux-Hugon [212] pointed out a considerable discrepancy between the values of the forbidden band width of $CdGeAs_2$ obtained from electrical and optical measurements. According to Leroux-Hugon, this is due to a greater complexity of the energy bands in $CdGeAs_2$ than in indium arsenide. In view of this, it is interesting to consider investigations of the band structures of $A^{II}B^{IV}C_2^V$ compounds carried out by Gashimzade [34, 217].

According to Gashimzade [34], the conduction band of these compounds is nonparabolic but spherically symmetrical. The valence band edge is similar to the valence band edges of semiconductors having the sphalerite-type lattices. Because of the lowering of the symmetry due to the transition from the sphalerite to the chalcopyrite structure, the triply degenerate (without allowance for spin) band of symmetry Γ_{15} (Parmenter's nomenclature [218] is used) becomes doubly degenerate. The band structure is similar to that found in wurtzite structures [226]. Gashimzade [34] concluded that the conduction bands of compounds with the sphalerite and the chalcopyrite structures are also similar, namely, the lower edge of the conduction band is in the center of the Brillouin zone and is nondegenerate.

When the forbidden band width is small, the interaction between the valence and conduction bands is strong and this results in the distortion of the conduction band. Such a distortion should reduce the effective carrier mass [204]. Moreover, the nonparabolicity of the conduction band should give rise to special properties, particularly temperature and carrier-density dependences of the effective mass, as found earlier for indium arsenide and antimonide [227, 228]. This effect has been discovered by Mamaev in $CdSnAs_2$.*

The effective mass of carriers in $CdSnAs_2$ was calculated by Gashimzade [34] using the value of the forbidden band width (0.26 eV [206]) taking into account the spin-orbit splitting, estimated from the spin-orbit splitting of the valence electrons of free atoms of elements forming the ternary compounds; these calculations yielded the value m* = $0.014m_0$.

The dispersion law for $CdSnAs_2$ can be expressed (making a number of assumptions) by means of a formula similar to that suggested by Kane [204], which predicts that the effective mass should increase with increasing carrier density and temperature. However, since the forbidden band width decreases when the temperature is increased, the effective carrier mass should decrease. These two tendencies are simultaneous and independent and, according to Gashimzade [34], may give rise to a minimum in the temperature dependence of the effective mass. He points out also that the valence band extrema are located on the circumferences of circles with centers at the point $\mathbf{k} = 0$, i.e., we can have an extremum loop, which was first described in [219].

The band structure of semiconducting crystals with the chalcopyrite structure is considered in [221].

We shall now discuss other properties of $A^{II}B^{IV}C_2^V$ compounds. Rud'† reports values of the properties of single crystals of ternary refractory compounds $ZnSiP_2$, $CdSiP_2$, and $ZnSiAs_2$.

*S. Mamaev, Dissertation for Candidate's Degree [in Russian], Gertsen State Pedagogical Institute, Leningrad (1962).

†Yu. V. Rud', Dissertation for Candidate's Degree [in Russian], Physicotechnical Institute of the Academy of Sciences of the USSR, Leningrad (1965).

TABLE 36. Lattice Parameters, Melting Points t_{mp}, Microhardness H, Density ρ, Effective Masses m^*/m_0, and Forbidden Band Widths ΔE of Some $A^{II}B^{IV}C_2^V$ Compounds [221, 229, 292]

Compound	Lattice parameters, nm (Å)		t_{mp}, °C	H, MN/m² (kgf/mm²)	ρ, g/cm³ (pycnometric)	m^*/m_0		ΔE, eV
	a	c				theoretical	experimental	
ZnSiP$_2$	0.540 (5.400)	1.0441 (10.441)	1500	10791 (1100)	3.35	0.096	0.08	2.3
CdSiP$_2$	0.5678 (5.678)	1.0431 (10.431)	1000	—	3.97	0.092	—	2.2
ZnGeP$_2$	—	—	1020	9614 (980)	—	—	—	2.2
CdGeP$_2$	—	—	776	8338.5 (850)	—	—	—	1.8
ZnSiAs$_2$	0.5606 (5.606)	1.089 (10.890)	1038	9025 (920)	4.69	0.071	—	1.6
ZnGeAs$_2$	—	—	850	6867 (700)	—	—	0.4	0.6
ZnSnAs$_2$	0.5851 (5.8515)	1.1703 (11.703)	775	4463 (455)	5.53	0.029	—	0.6
CdGeAs$_2$	0.5943 (5.9427)	1.12172 (11.2172)	665	4444 (453)	5.60	0.020	—	0.54
CdSnAs$_2$	—	—	615	3875 (395)	—	—	—	0.26

A characteristic feature of these three compounds is a wide forbidden band. In our opinion, the positions of elements forming these compounds in the periodic system suggest the possibility of high carrier mobilities. Measurements of the electrical conductivity and of the Hall effect show that the carrier mobility in ZnSiP$_2$ is 1000 cm$^2 \cdot$V$^{-1} \cdot$sec^{-1} for a carrier density of about 10^{17} cm^{-3}. The effective carrier mass in this compound is approximately 0.08m_0. Some properties of refractory $A^{II}B^{IV}C_2^V$ compounds are listed in Table 36.

Crystals of these refractory compounds were grown by Rud' using a method in which the synthesis is carried out at a relatively low temperature in a solvent with a low melting point, which need not be one of the components of the synthesized ternary compounds (synthesis from a molten solution); this method had not been tried before in the preparation of ternary semiconducting compounds. Crystals grown in the apparatus described in [55] from a solution in tin, antimony, or bismuth have the usual shape of an upright faceted prism, elongated along the [111] direction.

Detailed investigations of the electrical and photoelectric properties of ZnSiP$_2$ single crystals, prepared by the gas-transport reaction method (using iodine as the transport agent), are described in [81, 230]. The forbidden band width of this compound, deduced from optical measurements, decreases from 2.27 eV at 80°K to 2.22 eV at 295°K. The effective electron mass has been found to be $m^* = 0.4m_0$.

The photoelectric and electrical properties of single crystals of ZnSiAs$_2$, CdGeAs$_2$, and ZnSnAs$_2$ can be found in [231, 232]. The photoconductivity spectra of ZnSiAs$_2$ and CdGeAs$_2$, exhibiting p-type conduction, have been determined in the temperature range 80-300°K [231].

According to the results of the electrical measurements, the hole mobility in CdGeAs$_2$ is 45 cm$^2 \cdot$V$^{-1} \cdot$sec^{-1} when the hole density is $4 \cdot 10^{14}$ cm^{-3}.

The values of the forbidden band width of ZnSiAs$_2$ (2.10 eV at 295°K and 2.14 eV at 200°K, obtained from photoconductivity measurements [231]), are in agreement with the results of an investigation of the temperature dependence of the electrical conductivity. The activation energy of impurities in ZnSiAs$_2$, calculated from the temperature dependence of the short-circuit photocurrent, is 0.15 eV. However, the nature of the impurities responsible for this acceptor level has not been determined. The forbidden band width of CdGeAs$_2$ is 0.54 eV at 80°K and

TABLE 37. Thermal Expansion Coefficients α_L, Thermal
Conductivities $\varkappa$, Young's Moduli E, and Debye
Temperatures θ of Some $A^{II}B^{IV}C_2^V$ Compounds

Compound	$\alpha_L \times 10^6$, deg^{-1} [100]	$\varkappa$, mW·cm^{-1}·deg^{-1} (mcal·cm^{-1}·sec^{-1}·deg^{-1})		E, MN/cm^2 (Tdyn/cm^2) [100]	θ, °K [233]
		[212]	[100]		
ZnGeAs$_2$	1.0	114 (27)	66.5 (16)	12.9 (1.29)	—
ZnSnAs$_2$	2.3	74 (18)	58 (14)	9.8 (0.98)	271
CdGeAs$_2$	3.5	40 (9.6)	48 (11)	7.6 (0.76)	—
CdSnAs$_2$	4.7	70 (17)	40 (9.6)	7.5 (0.75)	234

0.50 eV at 295°K [231]. These values are somewhat lower than those obtained from the optical absorption spectra [225].

Masumoto and Isomura [232] prepared ZnSnAs$_2$ single crystals using the vertical Bridgman direction of crystallization method and graphite-lined quartz ampoules pulled at a rate of 0.5 mm/min; these crystals were grown after heating for two hours at a temperature exceeding the melting point by 45 deg. Masumoto and Isomura measured the lattice parameters, the melting point, and the sphalerite—chalcopyrite transition temperature. They also investigated the temperature dependences of the electrical conductivity, the Hall effect, the thermoelectric power, the optical transmission, and the optical reflection. To obtain a single crystal with the sphalerite structure, a crystal grown by the Bridgman method was annealed for 15 h at 720°C and then quenched in water. The chalcopyrite—sphalerite transition temperature is 632°C.

The published literature contains many examples of the considerable discrepancies between the experimental values of the melting points and the temperatures of phase transitions in $A^{II}B^{IV}C_2^V$ compounds [21, 201, 206, 212]. In view of this, precision determinations of these parameters are an urgent problem. Unfortunately, such determinations have not yet been carried out systematically. There is only one paper, that by Vaipolin and Korshak [42], dealing with the precision determination of the melting point and other thermal properties of CdSnAs$_2$. According to Vaipolin and Korshak [42], the melting point of this compound is 596 ± 2°C. The chalcopyrite—sphalerite transition temperature is 587°C. CdSnAs$_2$ crystals with the sphalerite structure, prepared by quenching from a high temperature and investigated by x-ray diffraction, have a lattice constant $a = 6.0510 \pm 0.0005$ Å, while CdSnAs$_2$ crystals with the chalcopyrite structure have $a = 6.0937$ Å and $c = 11.9184$ Å.

The thermal expansion, thermal conductivity, and elastic properties of crystals of some $A^{II}B^{IV}C_2^V$ compounds have been reported in [100, 211-213, 233]. Some of the results are given in Table 37.

The values of the Debye temperature have been deduced from measurements of the lattice thermal conductivity and of the specific heat near 4°K [233]. It is worth noting the discrepancy of the results obtained for the thermal conductivity as reported in [100, 211, 212] (cf. Tables 35 and 37). This discrepancy may be due to different synthesis methods and different subsequent heat treatments.

Unfortunately, the problem of the best synthesis conditions remains unsolved because it requires a detailed knowledge of the phase diagrams. Goryunova et al. [234] published an interesting paper on the problems of the kinetics of formation of ZnGeAs$_2$, which is one of the $A^{II}B^{IV}C_2^V$-type compounds. According to Goryunova et al. [234], two reactions take place:

TABLE 38. Electron Mobilities μ and Forbidden Band
Widths ΔE of $CdSnAs_2 - InAs$ Alloys

Alloy composition	$cm^2 \cdot V^{-1} \cdot sec^{-1}$	ΔE, eV
$CdSnAs_2$	1000	0.8
1 part $CdSnAs_2$ — 3 parts (2InAs)	750	0.6
1 part $CdSnAs_2$ — 1 part (2InAs)	600	0.2

at temperatures above the melting point of zinc (419°C), we have

$$Zn + 2As = ZnAs_2,$$

while at temperatures above the melting point of $ZnAs_2$ (771°C), the next stage produces the final compound:

$$ZnAs_2 + Ge = ZnGeAs_2.$$

Moreover, this reaction is accompanied by the dissociation of $ZnAs_2$, in accordance with the reaction

$$3ZnAs_2 = Zn_3As_2 + As_4.$$

At 1000°C, an equilibrium is reached:

$$ZnAs_2 (l) + Ge (l) \rightleftarrows ZnGeAs_2 (l).$$

Goryunova et al. [234] investigated the phase diagram of the $ZnGeAs_2 - 4Ge$ system and concluded that $ZnGeAs_2$ melts as a chemical compound, which is not dissociated in the solid state but does dissociate in the liquid phase.

We must mention that investigations of the temperature dependence of the electrical conductivity of $CdSnAs_2$ in the solid and liquid states [235] have demonstrated a constancy of the slope of the dependence $\ln \sigma = f(1/T)$ in both states, which indicates that this compound does not dissociate in the liquid phase.

In view of the unsolved problem of the synthesis of ternary compounds of the $A^{II}B^{IV}C_2^V$ type, we must regard as very important those investigations of solid solutions of quasibinary systems in which one of the components is an $A^{II}B^{IV}C_2^V$ compound.

In 1959, Folberth [13] reported the existence of solid solutions in $(Cd_{x/2}In_{1-x}Sn_{x/2})As$, $(Zn_{x/2}In_{1-x}Sn_{x/2})As$, $(Zn_{x/2}In_{1-x}Ge_{x/2})As$, and $(Zn_{x/2}Ga_{1-x}Ge_{x/2})As$ alloys in the range x = 0-1.

Fleischmann, Folberth, and Pfister [236] reported in 1959 the existence of solid solutions in alloys of the $A^I_{x/2} B^{IV}_{1-x} C^V_{x/2} D^{VI}$ type.

Some solid solutions, based on $CdSnAs_2$ and InAs, were synthesized by Mamaev [237], who demonstrated that in solid solutions of the $m(CdSnAs_2) - n(2InAs)$ system, the chalcopyrite structure is observed for m > n and the sphalerite structure is found for m < n. Mamaev synthesized alloys in evacuated and sealed quartz ampoules. He found that all his alloys were of n-type. Table 38 gives the values of the carrier mobility and the forbidden band width, obtained from the measurements of the temperature dependence of the electrical conductivity and Hall effect of some of the solid solutions prepared by Mamaev [237]. From the values given in Table 38, it follows that these solid solutions may have some practical applications.

According to Leroux-Hugon [239], the chalcopyrite structure in the $CdSnAs_2$–InAs system is found between 75 and 100% of $CdSnAs_2$.

Investigations of solid solutions in this system, as well as in the $CdSnSb_2$–InSb system, are reported also in a paper by Goryunova and Prochukhan [249]. The phase diagram of the $CdSnAs_2$–(2InAs) system has been plotted by Rupprecht and Maier [250], using the results of Goryunova and Prochukhan.

As already reported, $CdSnSb_2$ does not exist on its own. However, the existence of solid solutions with the sphalerite structure has been observed in alloys of the system $m(CdSnSb_2)$–$n(2InSb)$ in the range of concentrations $0 < m \leq n$ [249]. It is interesting to note that the same applies to the systems $ZnGeSb_2$–InSb and $ZnSnSb_2$–InSb [98, 251, 252] but the effect has not yet been explained satisfactorily from the physicochemical point of view.

There are reports in [239, 253, 254] of a detailed investigation of the electrical properties of solid solutions of $CdSnAs_2$–InAs system. The temperature dependences of the electrical conductivity and of the Hall effect in samples of the investigated alloys have been found to be typical of semiconductors with high impurity concentrations. The forbidden band widths of these solid solutions range from 0.2 to 1.4 eV and the values of the carrier mobility range from 400 to 1900 $cm^2 \cdot V^{-1} \cdot sec^{-1}$, which is of some practical interest. The carrier density in the investigated solid solutions ranges up to 10^{19} cm^{-3}. The results reported in these investigations suggest that when the synthesis and heat treatment conditions are selected in a suitable manner, a much higher carrier mobility could possibly be achieved in alloys of the $CdSnAs_2$–InAs system. According to Leroux-Hugon [239], the carrier mobility and the effective carrier mass of the $CdSnAs_2$–InAs system passes through a minimum near the composition consisting of 20% $CdSnAs_2$ and 80% InAs. The ratio of the carrier mobilities of opposite signs is close to unity throughout the full range of concentrations [253].

The problem of the existence of solid solutions in the $ZnSnAs_2$–InAs system is considered in many papers [38, 98, 250, 252, 255]. It is reported in [255] that a continuous series of solid solutions with the sphalerite structure is formed in this system and the lattice constant varies linearly with the concentration: from 0.585 nm (5.85 Å) for $ZnSnAs_2$ to 0.606 nm (6.06 Å) for InAs. It is reported in [98, 252] that all the solid solutions obtained are semiconductors; it is also mentioned that samples suitable for electrical measurements can be prepared by zone refining.

The $ZnGeAs_2$–InAs system has also been investigated [98, 250, 252, 255]. This system exhibits mutual solubility over the whole range of concentrations [252, 255]. Solutions with up to 80% $ZnGeAs_2$ have the sphalerite structure but at concentrations of $ZnGeAs_2$ higher than 85%, the chalcopyrite structure is observed with parameters ranging from $a = 0.573$ nm (5.73 Å) and $c = 1.126$ nm (11.26 Å) for 85% $ZnGeAs_2$–15% InAs to $a = 0.567$ nm (5.67 Å) and $c = 1.114$ nm (11.14 Å) for 100% $ZnGeAs_2$.

The solubility in the $ZnGeAs_2$–Ge system was investigated by Goryunova and Chiang Ping-Hsi [153]. They found a solid solution (homogeneity region) in $ZnGe_xAs_2$ in the range $1 < x \leq 4$; solid solutions with $x = 1.5$–4 have the sphalerite structure and those with $x = 1$–1.5 have the chalcopyrite structure.

Alloys based on $CdGeAs_2$ compounds have been investigated much less thoroughly. A eutectic phase diagram is shown in [250, 256] for the $2InAs$–$CdGeAs_2$ systems. Investigations of alloys based on $CdGeAs_2$, containing an excess of one of the components over the stoichiometric proportion, have been reported in [257]. The crystal structure of these alloys has been

TABLE 39. Density ρ, Melting (or Softening) Temperature T_{ph},
Microhardness H, Electrical Conductivity σ, and Forbidden
Band Width ΔE of $CdGeAs_2$ [126, 240, 258]

State	ρ, g/cm³	T_{ph}, °C	H, MN/m² (kgf/mm²)	σ, Ω^{-1} ·cm^{-1}	ΔE, eV electrical	ΔE, eV optical
Crystal	5.60	700	4611 (470)	10	0.53	0.6
Glass	5.35	410	3139 (320)	10^{-6}	1.1	0.6

found to be close to the chalcopyrite structure. Investigations carried out by Vaipolin, Osmanov, and Rud' [126] demonstrated that $CdGeAs_2$ has a considerable homogeneity (solid-solution) region in the solid state.

The mutual solubility of ternary $A^{II}B^{IV}C_2^V$ compounds has not yet been investigated systematically. We can mention only a study of Vaipolin, Osmanov, and Rud' [126] concerned with the phase diagram of the more complex system $ZnSnAs_2 - 2InAs - CdSnAs_2$.

The possible formation of glasses by $A^{II}B^{IV}C_2^V$ compounds is an interesting subject. These compounds crystallize in lattices with the tetrahedral coordination. Goryunova and Kolomiets [241] have pointed out that the covalent tetrahedral binding prevents the formation of glasses. This is in agreement with the observation that the amorphous state has not been observed in any elements of group IV with the diamond structure or in binary compounds with the sphalerite structure.

However, Goryunova's team reported the existence of $A^{II}B^{IV}C_2^V$ glasses [126, 240, 257, 285].* $CdGeAs_2$ and $CdGeP_2$ have been obtained in the amorphous state by quenching liquid phases in an aqueous solution of sodium chloride, during which the rate of cooling exceeded 200 deg/sec. As reported in [240], the presence of allotropic transitions can be observed most clearly in $CdGeAs_2$. This compound exhibits conchoidal fracture, optical uniformity, and diffuse maxima in the Debye diffraction pattern. The form of these maxima indicates retention of the tetrahedral short-range order in the glassy state. This is also supported by the slight change in the density observed on crystallization of the glass. It is reported in [126] and in Osmanov's dissertation that glasses are formed in the $Ge - CdAs_2$ system in a range extending to ±20 mol.% on both sides of $CdGeAs_2$.

Glassy $CdGeAs_2$ exhibits no intrinsic conduction in the 80-670°K range, which is in contrast to semiconducting glasses investigated earlier [259]. Measurements of the temperature dependence of the electrical conductivity of glassy $CdGeAs_2$ have established that heating above 405°C increases strongly the electrical conductivity and that the temperature dependence of the conductivity changes irreversibly, indicating a transition to the crystalline state near this temperature. Some results of investigations of the properties of $CdGeAs_2$ in the crystalline and glassy states are given in Table 39.

We must mention another important point. According to [224, 292], the semiconducting properties of $A^{II}B^{IV}C_2^V$ compounds are similar to those of the isoelectronic $A^{III}B^V$ compounds. The fact that the ternary compounds have somewhat narrower forbidden bands but higher values of the microhardness than the corresponding binary compounds indicates a stronger covalence of binding in the ternary compounds. This conclusion is very important in consideration of any practical applications of $A^{II}B^{IV}C_2^V$ compounds.

---

*See also É. O. Osmanov, Dissertation for Candidate's Degree [in Russian], Silicate Chemistry Institute, Leningrad (1965).

CHAPTER V

OTHER TYPES OF TERNARY COMPOUNDS

Defect Ternary Compounds

As demonstrated in Chapter II, ternary defect phases can be regarded as combinations of simpler normal and defect phases and, in the case of phases with a similar crystal structure, superstructure-type ternary compounds can be formed in systems of solid solutions.

In many cases, when we consider the possibility of the formation of ternary compounds with the tetrahedral coordination of atoms in some ternary system, we are forced to use the term "ternary phase" because of the absence of information on the phase diagram of the ternary system considered, or even on the nature of the melting of the postulated ternary compound and its homogeneity range. Thus, a ternary phase $A_3^{III}B^V C_3^{VI}$, predicted by Goryunova [12], may represent simply a solid solution of $A^{III}B^V$ and $A_2^{III}C_3^{VI}$ compounds. In order to determine whether a given ternary phase is a solid solution or a chemical compound, an additional investigation is required in each specific case.

Table 40 lists known ternary defect phases and compounds together with their components. It has been reported [148, 242, 260] that phases Nos. 2-5 are formed in systems of solid solutions, which is in good agreement with the point of view put forward in the preceding para-

TABLE 40. Known Ternary Defect Phases and Compounds
Forming These Phases

Phase No.	Average no. of valence electrons per atoms	Ternary phase	Binary compounds
1	4.57	$A_2^I B^{II} C_4^{VII}$ (Ag$_2$HgI$_4$)	$2A^I C^{VII} + B^{II} C_2^{VII}$ (2AgI+HgI$_2$)
2	4.57	$A^{II} B_2^{III} C_4^{VI}$ (ZnGa$_2$S$_4$)	$A^{II} C^{VI} + B_2^{III} C_3^{VI}$ (ZnS+Ga$_2$S$_3$)
3	4.57	$A_2^{II} B^{IV} C_4^{VI}$ (Zn$_2$GeS$_4$)	$2A^{II} C^{VI} + B^{IV} C_2^{VI}$ (2ZnS+GeS$_2$)
4	4.80	$A^{II} B^{IV} C_3^{VI}$ (ZnGeS$_3$)	$A^{II} C^{IV} + B^{IV} C_2^{VI}$ (ZnS+GeS$_2$)
5	4.57	$A_3^{III} B^V C_3^{VI}$ (Ga$_3$PSe$_3$)	$A^{III} B^V + A_2^{III} C_3^{VI}$ (GaP+Ga$_2$Se$_3$)
6	5.33	$A^{III} B^V C_4^{VI*}$	$A_2^{III} C_3^{VI} + B_2^V C_5^{VI}$

*Does not crystallize in diamond-like structure (A^{III} = B, Al; B^V = P, As; C^{VI} = S, Se, Te).

79

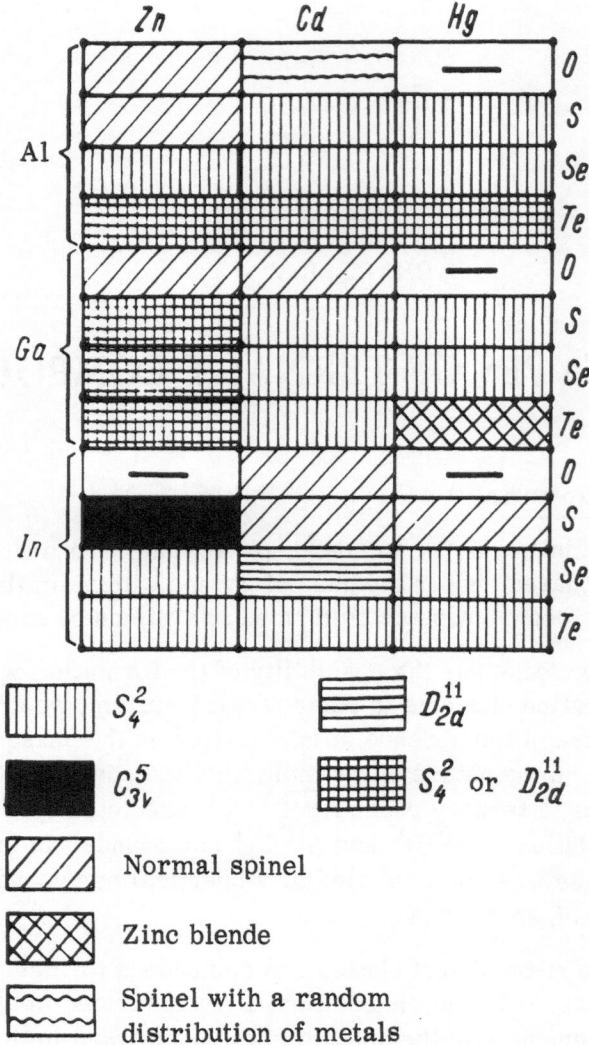

Fig. 36. Structures of $A^{II}B_2^{III}C_4^{VI}$ compounds.

graph. Of all the defect ternary compounds listed in Table 35, only those of the $A^{II}B_2^{III}C_4^{VI}$ and $A_2^I B^{II}C_4^{VII}$ type have been investigated in some detail and we shall discuss them now in the next two subsections.

$A^{II}B_2^{III}C_4^{VI}$ Compounds

$A^{II}B_2^{III}C_4^{VI}$ compounds, formed in $A^{II}C^{VI}-B_2^{III}C_3^{VI}$ systems of solid solutions, have been discovered and investigated by many workers [243-245, 260, 290, 291]. It has been established that compounds corresponding to the $A^{II}B_2^{III}C_4^{VI}$ composition are formed in such solid solution systems when the ratio of the binary components is 1:1. Even in those cases when the binary components are not cubic, as is the case in the $CdSe-In_2Se_3$ system, the ternary compounds ($CdIn_2Se_4$ in this case) can have the cubic structure.

Using classifications of semiconducting compounds, suggested by various authors, we can classify $A^{II}B_2^{III}C_4^{VI}$ group compounds as derivatives of $A^{II}B^{VI}$ binary compounds with substituent atoms whose valences are two, three, and three [5]. When the substituent atoms are located at tetrahedral and octahedral vacancies, mixed binding (tetrahedral and octahedral) of the type observed in spinels, is observed. The cause of tetrahedral or mixed binding will be considered later.

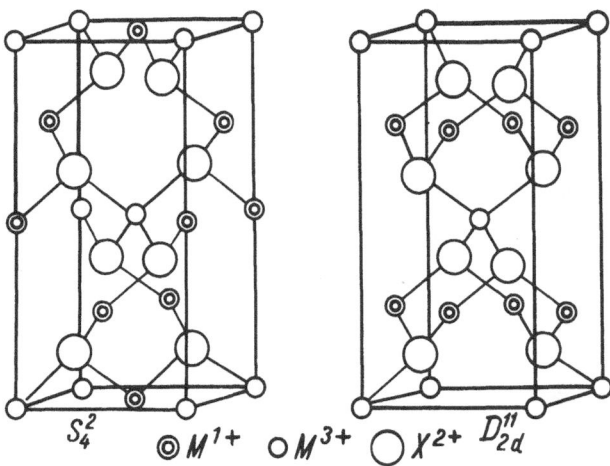

Fig. 37. Structures with the S_4^2 and D_{2d}^{11} space symmetries.

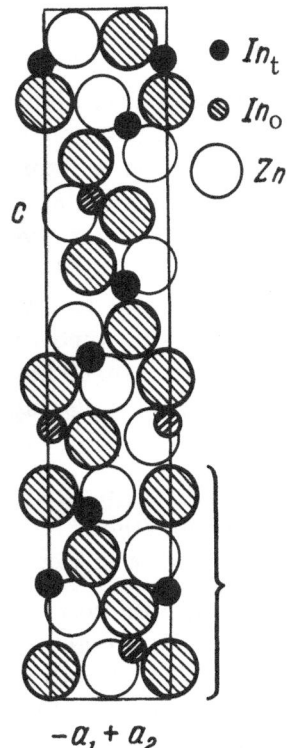

Fig. 38. Structure of $ZnIn_2S_4$.

The structure of $A^{II}B_2^{III}C_4^{VI}$ compounds can be deduced by successive substitutions in the zinc-blende structure. By doubling the sphalerite structure and replacing two A^{II} atoms with A^I and A^{III} atoms, we obtain the chalcopyrite cell; by doubling the chalcopyrite cell and replacing four A^I atoms with two A^{II} atoms, we obtain the $A^{II}B_2^{III}C_4^{VI}$ cell. The structure deduced in this way differs from that of chalcopyrite only by the presence of two vacant metal sites [246, 247].

A most detailed x-ray diffraction investigation of $A^{II}B_2^{III}C_4^{VI}$ compounds was carried out by Hahn and his colleagues [243, 248, 260]. They established that these compounds crystallize mainly in two structures: spinel (unit cell shown in Fig. 38) or tetrahedral structure with the space group S_4^2 or D_{2d}^{11}. The formation of a particular structure depends on the tendency to form the tetrahedral coordination around a group II or III metal anion. In the structure of a normal spinel, atoms of group II occupy tetrahedral vacancies and those of group III occupy octahedral vacancies. The tendency of trivalent metals to form the tetrahedral coordination around oxygen and sulfur atoms is weaker than the corresponding tendency in relation to selenium and tellurium atoms. Therefore, spinel structures are observed only in compounds based on oxygen and sulfur, while selenides and tellurides crystallize in tetrahedral structures having the space groups S_4^2 or D_{2d}^{11}..

Figure 36 shows the structures of $A^{II}B_2^{III}C_4^{VI}$ compounds. In some cases, in view of the similarity of the intensity of scattering of atoms of groups II and III, it is not possible to establish reliably which of the two tetrahedral structures is exhibited by a given compound. The structures corresponding to the space groups S_4^2 and D_{2d}^{11} are similar and, as shown in Fig. 37, they differ only in one respect: in the D_{2d}^{11} case, both metals occupy the tetrahedral vacancies in the form of layers, while in the S_4^2 structure, atoms are distributed in a less ordered manner, which is very similar to the chalcopyrite structure.

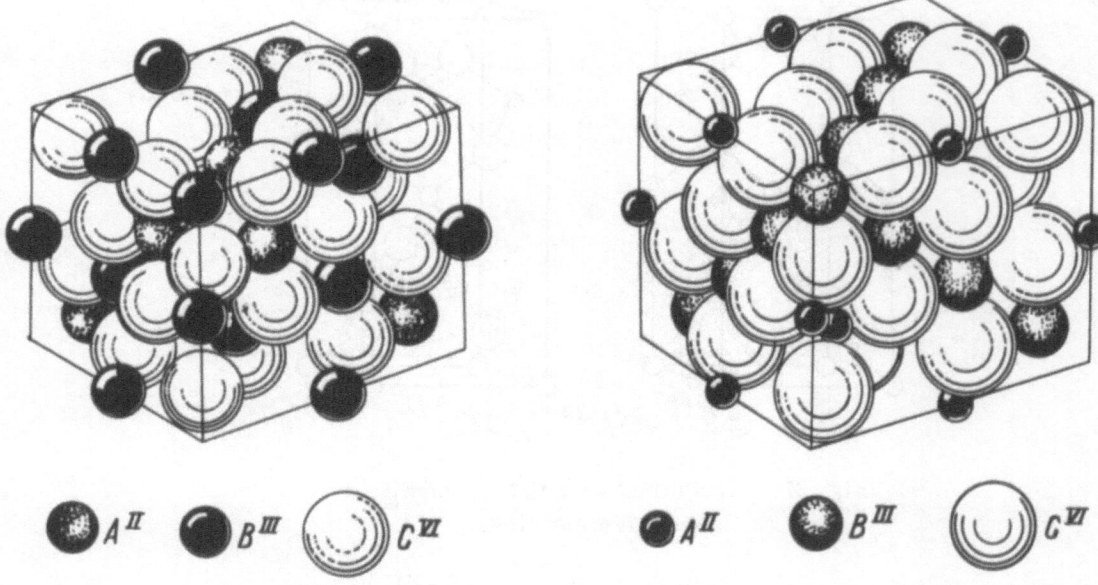

Fig. 39. Structure of an inverse spinel.   Fig. 40. Structure of a normal spinel.

Among the sulfides, the structure of $ZnIn_2S_4$ was not determined in the early investigations, but was eventually reported in [261]. It has been found that $ZnIn_2S_4$ forms a hexagonal-rhombohedral cell belonging to a space group $R3m-C_{3v}^5$ with the following lattice parameters: $a = 0.385$ nm (3.85 Å) and $c = 3.706$ nm (37.06 Å). The unit cell of this compound consists of twelve close-packed layers of sulfur atoms whose sequence along the c-axis direction is given by ABCACABCBCAB... . The octahedral vacancies contain only half of the total number of indium atoms, while the other half, together with zinc atoms, occupies the tetrahedral vacancies. The structure of $ZnIn_2S_4$ is shown in Fig. 38. If we approach this structure from the point of view of the hexagonal system, we find that the unit cell splits into three blocks, arranged along the c axis and bound by weak sulfur–sulfur bonds. For this reason, $ZnIn_2S_4$ has the layered structure and can be easily split into very thin plates.

Figure 36 shows the structures of compounds formed by elements of the B subgroups of the periodic system. However, many compounds of this type are known in which other elements are found. In such cases, we obtain the structure of an inverse spinel, shown in Fig. 39.

As already reported, according to Hahn [260], the thiogallate structure with the tetrahedral coordination of atoms appears only where there exists a strong tendency of metal atoms to form the tetrahedral coordination around an anion and a strong polarizing effect on the anion by the metal atoms. When these tendencies are weak, divalent atoms form spinel structures. When a divalent atom has a stronger tendency to form the tetrahedral coordination than the corresponding tendency of a trivalent atom, we obtain the normal spinel structure (Fig. 40), while in the reverse case we obtain the inverse spinel structure. The structure of an inverse spinel can be deduced from that of a normal spinel by interchanging divalent atoms and one half of the trivalent atoms.

In other words, in an inverse spinel, the tetrahedral vacancies contain half the total number of trivalent atoms while the other half and divalent atoms are located in the octahedral vacancies. $MgIn_2S_4$, $FeIn_2S_4$, $CoIn_2S_4$, and $NiIn_2S_4$, described in [243], have the inverse spinel structure, which can be explained on the basis of considerations just given. Mooser and Pearson [9] considered chemical binding in semiconductors in general and predicted that these compounds should exhibit semiconducting properties.

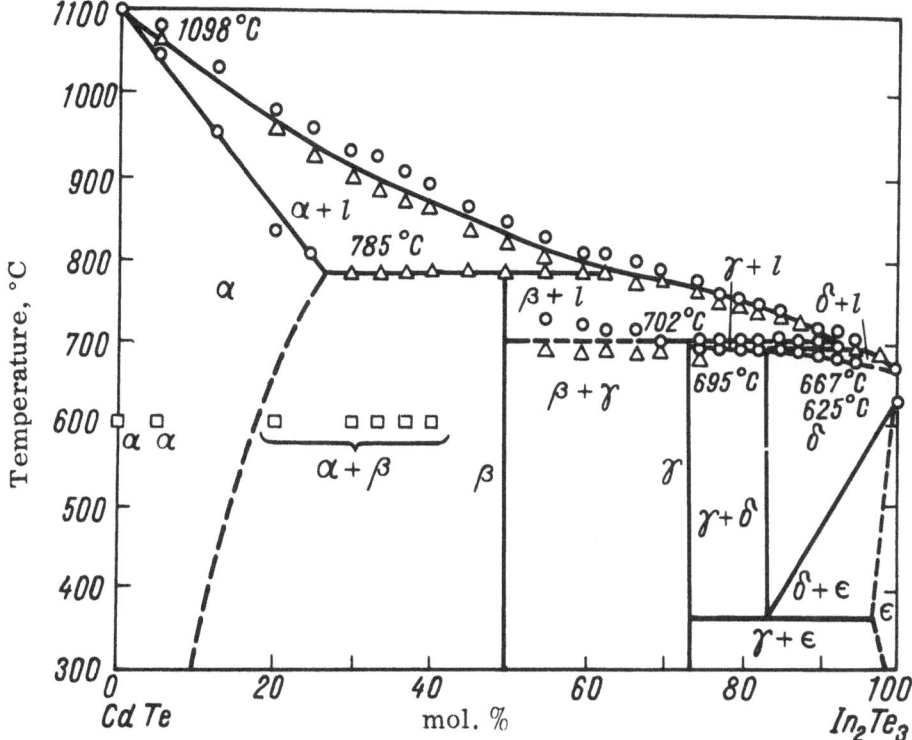

Fig. 41. Phase diagram of the pseudobinary system $CdTe-In_2Te_3$.

It should be pointed out that when transition elements participate in the formation of ternary compounds, we can also obtain the normal spinel structure. This structure is exhibited, for example, by $ZnFe_2O_4$, MgV_2O_4, $ZnCo_2O_4$, and other compounds. The stability of such compounds results from the covalent type of binding [14], which is due to the fact that, for example, in $ZnCo_2O_4$, electrons of zinc and oxygen form the sp^3 hybrid bonds and cobalt electrons form the d^2sp^3 hybrid bonds.

Semiconducting properties appear mainly in those oxygen compounds in which at least one of the elements is a transition metal. These compounds have a conductivity higher than that of dielectrics and lower than that of most other semiconductors (10^{-7}-10^{-11} $\Omega^{-1} \cdot cm^{-1}$). It is mentioned in [262] that the synthesis of solid solutions on the basis of poorly conducting $MgCr_2O_4$ and $ZnFe_2O_4$ and highly conducting magnetite (which has exceptional properties) yields semiconducting materials with resistivities within the range 10^{-2}-10^{10} $\Omega \cdot cm$.

Compounds of the $A^{II}Fe_2O_4$ type, which have the spinel structure, belong to a class of semiconducting ferrites [263]. The divalent element in these compounds can be copper, magnesium, zinc, cadmium, iron, cobalt, or nickel. However, ferrites are outside the scope of substances considered in the present monograph. All sulfides, selenides, and tellurides investigated up to now are typical semiconductors. We shall consider only the properties of sulfides, selenides, and tellurides formed with elements of B subgroups.

Various methods are currently used to prepare these compounds. Many compounds, first synthesized by Hahn and his colleagues [260], were prepared from a mixture of binary compounds taken in the required proportions. Samples of the binary compounds were carefully ground, mixed, pressed into tablets, and then fired at suitable temperatures. Naturally, this technique did not produce single-crystal samples. Further investigations of these compounds was impeded by two factors: the high melting points of the sulfides and the incongruent nature of the melting of some selenides and tellurides. The relatively low melting points of the selen-

TABLE 41. Color, Shape, and Dimensions of Crystals
of $A^{II}B_2^{III}C_4^{VI}$ Compounds

Compound	Color	Shape	Dimensions, mm
$ZnGa_2S_4$	Colorless	Polyhedra	$0.5 \times 0.5 \times 0.5$
$CdGa_2S_4$	Colorless	Columns	$6 \times 2 \times 2$
$HgGa_2S_4$	Yellow	Needles	$8 \times 0.3 \times 0.3$
$ZnIn_2S_4$	Yellow	Plates	$10 \times 10 \times 0.1$
$CdIn_2S_4$	Red	Octahedra	$5 \times 5 \times 5$
$HgIn_2S_4$	Black	Octahedra	$1 \times 1 \times 1$
$ZnGa_2Se_4$	Orange-red	Polyhedra	$0.5 \times 0.5 \times 0.1$
$CdGa_2Se_4$	Orange-red	Columns	$1 \times 1 \times 4$
$HgGa_2Se_4$	Black	Needles	$0.5 \times 0.5 \times 10$
$ZnIn_2Se_4$	Dark red	Columns	$1 \times 1 \times 5$
$CdIn_2Se_4$	Black	Columns	$1 \times 1 \times 5$

ides and tellurides have made it possible to use methods of direct fusion of components, followed by recrystallization [301], but it has been found [265] that $HgIn_2Te_4$ is the only compound among indium selenides and tellurides which melts congruently. Therefore, in the zone recrystallization of these compounds, it is necessary to use the special method described in Chapter III. Figure 41 shows the phase diagram of the $CdTe-In_2Te_3$ system, which shows that the compound $CdIn_2Te_3$ decomposes on melting. In view of this, one must view with suspicion the suggestion made in [247] that selenides and tellurides can be prepared by the standard method of growth of crystals and zone recrystallization.

The melting points of the ternary sulfides have not yet been determined but it is known that they are very high and are likely to be intermediate between the melting points of the binary sulfides which are components of a given ternary sulfide. Moreover, there are grounds for assuming that the equilibrium vapor pressure of sulfur above molten sulfides is very high. It has been reported in [266] that to synthesize $CdIn_2S_4$, a sulfur vapor pressure of 5 atm above a mixture of CdS and In_2S_3 was necessary at a temperature of 1150°C, which is considerably below the melting points of CdS. It follows that the most suitable method for the preparation of single crystals of sulfides is the method of gas-transport reactions described in Chapter III.

It has been demonstrated [63, 68, 74-76, 267] that this method can be used to prepare single crystals of various compounds. Table 41, which is based on the data reported in the cited papers, lists the dimensions, color, and shape of the obtained crystals.

The crystal lattice parameters of $A^{II}B_2^{III}C_4^{VI}$ compounds as well as their densities, obtained experimentally and calculated from x-ray diffraction data, have been known for some time and are given in [12, 302]; they will not be repeated in the present monograph.

TABLE 42. Properties of $A^{II}B_2^{III}C_4^{VI}$ Compounds

Compound	Forbidden band width ΔE, eV	ρ illum. $\Omega \cdot cm$	ρ dark, $\Omega \cdot cm$	Wavelength, μ	
				λ_{max}	λ_{edge}
$ZnGa_2S_4$	—	$1.5 \cdot 10^{10}$	$4 \cdot 10^{11}$	0.39	0.36
$CdGa_2S_4$	3.44	$2.5 \cdot 10^8$	$8 \cdot 10^{13}$	0.35	0.358
$HgGa_2S_4$	2.84	$7 \cdot 10^4$	$1 \cdot 10^{10}$	0.49	0.433
$ZnIn_2S_4$	2.6—2.86	$2 \cdot 10^8$	$1 \cdot 10^{14}$	0.48	0.47
$CdIn_2S_4$	2.18—2.3	$3 \cdot 10^4$	$1 \cdot 10^8$	0.54	0.54
$HgIn_2S_4$	2.0	$1.8 \cdot 10^6$	$2.7 \cdot 10^6$	0.62	0.62
$ZnGa_2Se_4$	—	$1 \cdot 10^8$	$2.5 \cdot 10^{12}$	0.57	—
$CdGa_2Se_4$	2.43	$1.1 \cdot 10^5$	$4 \cdot 10^{11}$	0.48	0.515
$HgGa_2Se_4$	1.95	$2.7 \cdot 10^4$	$1.4 \cdot 10^7$	0.62	0.677
$ZnIn_2Se_4$	1.82—1.95—2.6	$1.8 \cdot 10^4$	$2.4 \cdot 10^7$	0.58	0.683
$CdIn_2Se_4$	1.45—1.72	$7 \cdot 10^5$	$8 \cdot 10^5$	0.77	0.730
$HgIn_2Se_4$	0.6	—	—	—	—
$ZnIn_2Te_4$	1.4—1.8	$1.7 \cdot 10^4$—$1 \cdot 10^{8*}$		—	—
$CdIn_2Te_4$	0.88—0.9	$2 \cdot 10^{5*}$		—	—

*Determined from measurements of the electrical conductivity carried out under normal conditions at room temperature.

Practically nothing is known about the microhardness of $A^{II}B_2^{III}C_4^{VI}$ compounds. One report [135] states that the microhardness of solid solutions of the $CdIn_2Se_4$—$CdIn_2Te_4$ system ranges from 3 GN/m² (300 kgf/mm²) for $CdIn_2Se_4$ to 2.2 GN/m² (220 kgf/mm²) for $CdIn_2Te_4$.

Some physical properties of $A^{II}B_2^{III}C_4^{VI}$ compounds have been investigated fairly thoroughly. The forbidden band widths of these compounds vary within a very wide range. Table 42 lists the values of the forbidden band width, determined by different investigators using various methods. In some cases, the discrepancies between the results are considerable, particularly in the case of $ZnIn_2Se_4$. It is possible that these discrepancies are due to the incongruent nature of melting of $ZnIn_2Se_4$ and different methods used to synthesize this compound. The values of the forbidden band widths of $ZnGa_2S_4$ and $ZnGa_2Se_4$ are not given because crystals of dimensions sufficient for measurements of the forbidden band width have not yet been obtained for these

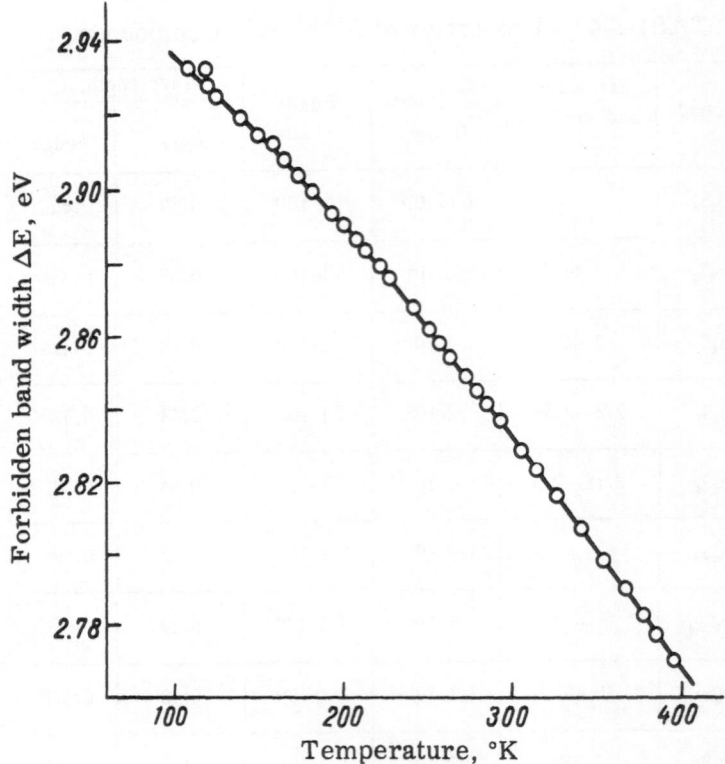

Fig. 42. Temperature dependence of the forbidden band
width of $ZnIn_2S_4$.

compounds. It has been reported in [269] that the forbidden band widths of ternary indium chalcogenides are equal to half the sums of the values of the forbidden bands of their binary components. However, the calculation of the forbidden bands in this way does not yield satisfactory
agreement with the experimentally obtained values for compounds containing gallium and cannot be employed even in approximate estimates of the unknown forbidden band widths of $ZnGa_2S_4$
and $ZnGa_2Se_4$. Figure 42 shows the temperature dependence of the forbidden band width of
$ZnIn_2S_4$ [267].

Investigations of the optical properties of some ternary chalcogenides have yielded dependences of the absorption coefficient on the photon energy. Comparison of these dependences
with various theoretical formulas has shown no obvious evidence of indirect transitions [74].

The most interesting property of ternary chalcogenides is their photosensitivity. The
photoconductivity of these compounds has been investigated by many workers [74, 266, 268,
270]. These investigations have been carried out on crystals with contacts made of indium
amalgam (in the case of compounds containing indium) or gallium amalgam (in the case of compounds containing gallium). The dark current has been measured as a function of an applied
electric field, and the photocurrent as a function of the applied field and the wavelength of the
incident light. No dependence of the photocurrent on the orientation of these crystals has been
found. $Me^{II}Ga_2Se_4$ compounds exhibit a linear dependence of the photocurrent on the illumination intensity, while $Me^{II}Ga_2S_4$ have a dependence which is close to the square root law. In all
these compounds, the dependence of the photocurrent on the voltage is strictly linear right up
to breakdown, which occurs (in the majority of these compounds) in fields stronger than
10^4 V/cm. Figures 43 and 44 show the photoconductivity spectra of some of the investigated compounds. Table 42 lists the values of ρ_{illum}, which is the resistivity during illumination of the
order of 10^{14} photons/sec of λ_{max} wavelength (λ_{max} is the wavelength of the sensitivity max-

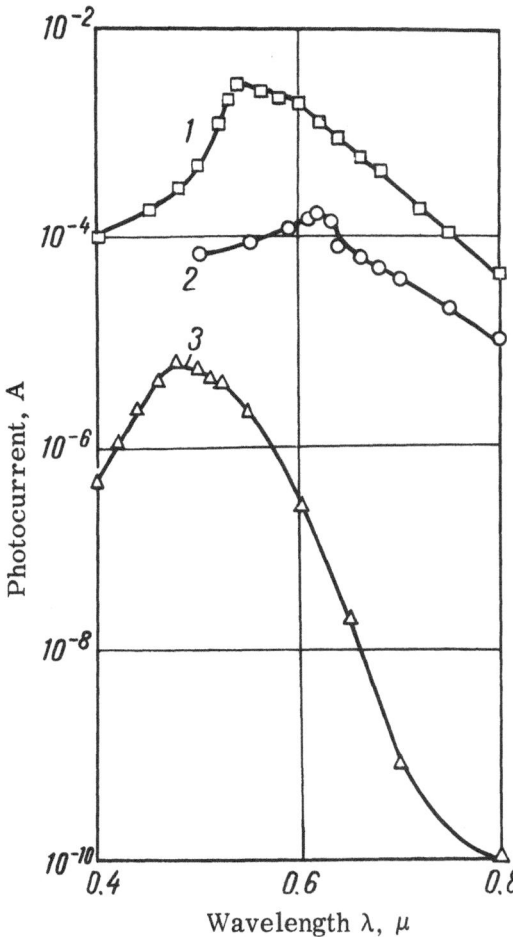

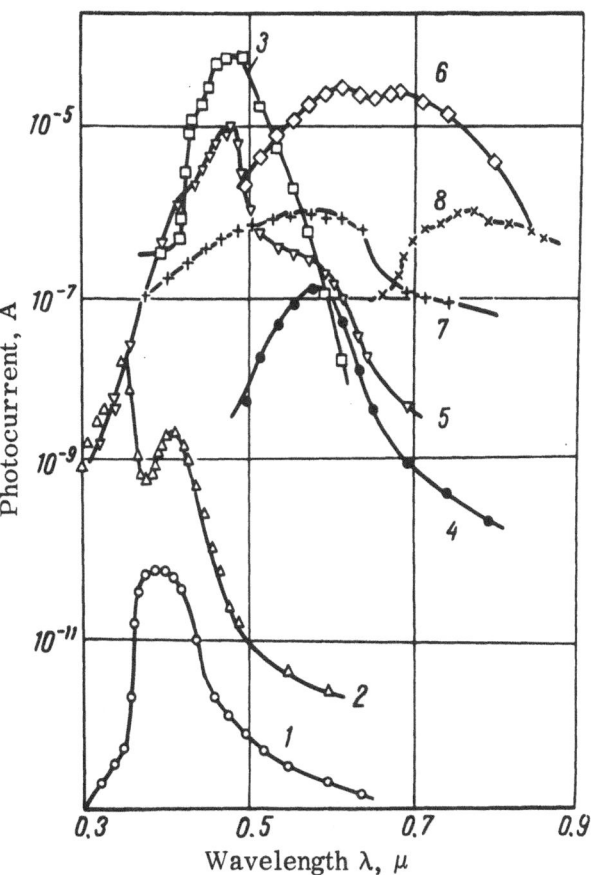

Fig. 43. Photocurrent spectra of some $A^{II}B_2^{III}C_4^{VI}$ sulfides: 1) $HgIn_2S_4$ (E = 300 V/cm); 2) $CdIn_2S_4$ (E = 120 V/cm); 3) $ZnIn_2S_4$ (E = 1000 V/cm).

Fig. 44. Photocurrent spectra of some $A^{II}B_2^{III}C_4^{VI}$ sulfides and selenides: 1)$ZnGa_2S_4$; 2) $CdGa_2S_4$; 3) $HgGa_2S_4$; 4) $ZnGa_2S_4$; 5) $CdGa_2Se_4$; 6) $HgGa_2Se_4$; 7) $ZnIn_2Se_4$; 8) $CdIn_2Se_4$.

imum), and the values of ρ_{dark}, which is the resistivity in darkness. It follows from Table 42 that the most promising photoconductor is $CdGa_2Se_4$, which has the highest ratio ρ_{illum}/ρ_{dark}. The relatively high dark conductivity of some of the crystals is likely to be due to the presence of a large number of impurities. Moreover, it has been found [315] that the conductivity of $CdIn_2Se_4$ samples changes by more than five orders of magnitude when they are annealed in selenium vapor. Similarly, the dark resistivity of $CdIn_2S_4$ depends very strongly on the synthesis method: samples, $4 \times 2 \times 0.5$ mm, prepared from the melt in vacuum, have a resistance of 1 Ω, whereas similar samples prepared under a sulfur vapor pressure amounting to 490 kN/m^2 (5 atm), have resistances ranging from 10 to 100 MΩ [266]. The activation of these samples with copper or gold increases their photosensitivity by three orders of magnitude. The monovalent atoms of copper and gold occupy positions of the divalent atoms of cadmium and are compensated by equal numbers of the trivalent atoms of indium at the cadmium sites. The optimum molar concentration of the activator is 2×10^{-3}.

Ternary chalcogenides are photoconducting phosphors and they exhibit luminescence. The luminescence spectrum of $CdIn_2S_4$ is known to have two maxima, at 1.42 and 1.65 eV [271]. The greenish yellow luminescence of $CdGa_2S_4$ reaches its maximum intensity at 2.15 eV

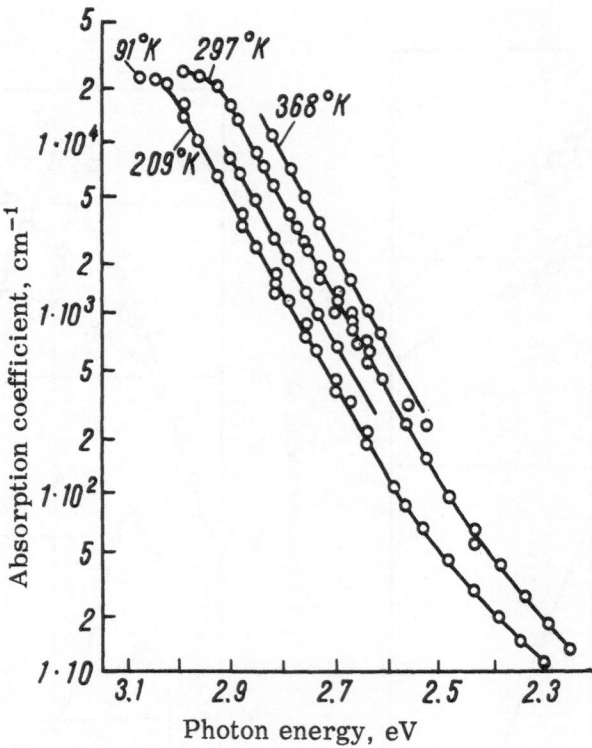

Fig. 45. Dependence of the absorption coefficient
of $ZnIn_2S_4$ on the photon energy.

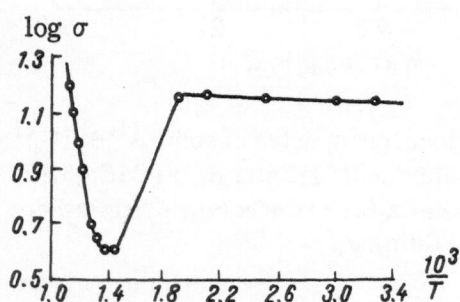

Fig. 46. Temperature dependence
on the conductivity of $CdIn_2Se_4$.

and this intensity increases when the temperature is lowered. These two compounds also exhibit thermoluminescence when they are cooled in darkness to 60°K and heated at a rate of 1 deg/sec. The intensity of the thermoluminescence of $CdGa_2S_4$ is higher than that of $CdIn_2S_4$. It is reported in [271] that the luminescence centers in these two compounds are complexes formed by anions around a cation vacancy. When the nature of the anion or the cation is altered, there should be a corresponding change in luminescence. Such a change can explain why it is necessary to subject these two compounds to annealing in sulfur vapor if they are prepared from the melt.

The luminescence of $ZnIn_2S_4$ has also been investigated. Identical results have been obtained in two different investigations [75, 76]. $ZnIn_2S_4$, excited with ultraviolet radiation, exhibits red luminescence. A stronger luminescence is observed for samples doped with copper. The luminescence spectra of doped and undoped $ZnIn_2S_4$ have been determined. At room temperature, both types of sample exhibit a peak at 1.58 eV, but at low temperatures the spectra of doped and undoped samples become different.

The dependence of the absorption coefficient of $ZnIn_2S_4$ on the photon energy has also been determined (Fig. 45). The optical absorption and the absorption coefficient K have been determined also for other compounds. Table 42 lists the values of the wavelength λ_{edge}, at which $K = 500$ cm^{-1}.

The optical properties of compounds containing heavy elements have been investigated to a much lesser degree. The optical properties of $CdIn_2Te_4$ have been studied at room temperature in the wavelength range 1-37 μ [316]. Values of absorption coefficient have been determined for n-type samples with various carrier densities. Investigations of the dependences of the absorption coefficient on the photon energy have indicated that direct and indirect transitions take place in this compound.

The electrical conductivity, thermoelectric power, and rectification of several indium—zinc and indium—cadmium selenides and tellurides are reported in [301]. Measurements of the thermoelectric power and its temperature dependence indicate that all the investigated compounds (with the exception of $ZnIn_2Te_4$) are p-type semiconductors which exhibit a change of the sign of the thermoelectric power in the temperature range 90-120°C. Measurements of the value of the thermoelectric power (in $\mu V/deg$) have given the following results:

$$ZnIn_2Se_4 \ldots \ldots \ldots \ldots \ 1000$$
$$CdIn_2Se_4 \ldots \ldots \ldots \ldots \ \ 70$$
$$ZnIn_2Te_4 \ldots \ldots \ldots \ldots \ 380$$
$$CdIn_2Te_4 \ldots \ldots \ldots \ldots \ 580$$

Investigations of the rectification at a point contact made of phosphor bronze or tungsten have yielded positive results only for the two selenides given in the above table. The absence of the rectification effect in the two tellurides may be due to the fact that a suitable etchant has not been used for these materials [301].

The temperature dependence of the conductivity of $CdIn_2Se_4$ is the same as that of $A^{III}B^V$ compounds. It is evident from Fig. 46 that at temperatures exceeding 450°C, this ternary compound exhibits an intrinsic conduction region.

Carrier mobilities in ternary chalcogenides are low and usually do not exceed several tens of $cm^2 \cdot V^{-1} \cdot sec^{-1}$. However, there are grounds for assuming that when the purity and perfection of single crystals are improved, the carrier mobility can reach much higher values. It has been pointed out in [317] that the mobility in complex ternary compounds is at least of the same order as the mobility in the binary compounds of which these ternary compounds are formed. Thus, in n-type single crystals of $HgIn_2Te_4$, prepared by the Bridgman method and having a carrier density of $3.5 \cdot 10^{15}$ cm^{-3}, the mobility is 200 $cm^2 \cdot V^{-1} \cdot sec^{-1}$ [272], whereas in n-type $CdIn_2Te_4$ with a carrier density of 10^{14} cm^{-3}, the mobility can reach 4000 $cm^2 \cdot V^{-1} \cdot sec^{-1}$ [39].

Thus, the properties of $A^{II}B_2^{III}C_4^{VI}$ compounds vary over a wide range of values which should make them useful in various applications.

$A_2^I B^{II} C_4^{VII}$ Compounds

Only two $A_2^I B^{II} C_4^{VII}$ compounds are known: Cu_2HgI_4 and Ag_2HgI_4; they can be obtained from the corresponding $A^I B^{VII}$ compounds when the valences of the substituent atoms are one, one, and two [5]. We have mentioned already that when the substituent atoms have higher valences, defect structures are formed. In $A_2^I B^{II} C_4^{VII}$ compounds, one quarter of the metal sites is vacant. It is evident from Table 40 that $A_2^I B^{II} C_4^{VII}$ compounds dissociate into simpler compounds: $A^I B^{VII}$ and $A^{II} C_2^{VII}$. Thus, for example, Cu_2HgI_4 can be dissociated into CuI, which has the zinc-blende structure, and HgI_2, which has a defect structure (this structure can be deduced from the sphalerite lattice when one quarter of the tetrahedral vacancies is filled with mercury atoms).

Various methods can be used to prepare Cu_2HgI_4 and Ag_2HgI_4. For example, Cu_2HgI_4 can be prepared chemically from solutions by reaction between K_2HgI_4 and $CuSO_4$. We can use also synthesis, based on the reaction between ground powders of, for example, HgI_2 and AgI, but in the majority of cases the samples prepared in this way contain excess HgI_2. Hahn and his col-

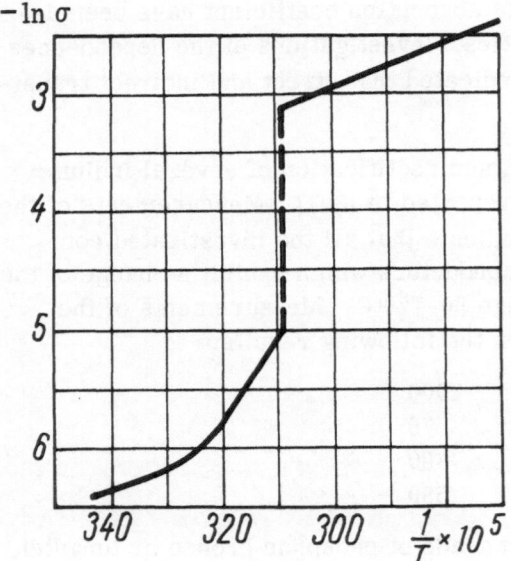

Fig. 47. Temperature dependence of the
electrical conductivity of Ag_2HgI_4.

leagues have been able to prepare $A_2^I B^{II} C_4^{VII}$
compounds by the fusion of stoichiometric proportions of elements in evacuated ampoules [273].

It is reported in [274] that single crystals
of Ag_2HgI_4 can be prepared by a suitable experimental technique.

X-ray diffraction investigations of $A_2^I B^{II} C_4^{VI}$
compounds have established that they have two
modifications: a high-temperature α modification and a low-temperature β modification.
Ketelaar [275] has found that the high-temperature modifications of Ag_2HgI_4 and Cu_2HgI_4 have
structures similar to that of zinc blende. Vacancies in the metal lattice are distributed at random. At low temperatures, the β modification
becomes stable; it has a tetragonal structure with
a more ordered distribution of atoms. According to Ketelaar's observations, based on changes
in color, the $\beta \rightleftharpoons \alpha$ transition is probably continuous. In other words, when the temperature
is increased, a gradual disordering of atoms takes place and each of these two modifications
contains both types of structure. The transition temperatures are 70°C for Cu_2HgI_4 and 50.7°C
for Ag_2HgI_4.

Ketalaar has determined the lattice parameters of the low-temperature (β) and high-
temperature (α) modifications of these compounds. The high-temperature modifications crystallize in structures similar to that of zinc blende and have the following lattice parameters:
0.6383 nm (6.383 Å) for Ag_2HgI_4 at 60°C and 0.6103 nm (6.103 Å) for Cu_2HgI_4 at 90°C. It is interesting to note that these parameters are similar to the parameters of the γ modifications of
AgI and CuI, which are 0.6498 nm (6.498 Å) and 0.6049 nm (6.049 Å), respectively.

In the determination of the structure of the low-temperature modifications, Ketelaar
has not interpreted some of the interference bands in the Debye diffraction patterns. These
bands have been interpreted by others [273, 276]. It has been found that β-Cu_2HgI_4 and β-

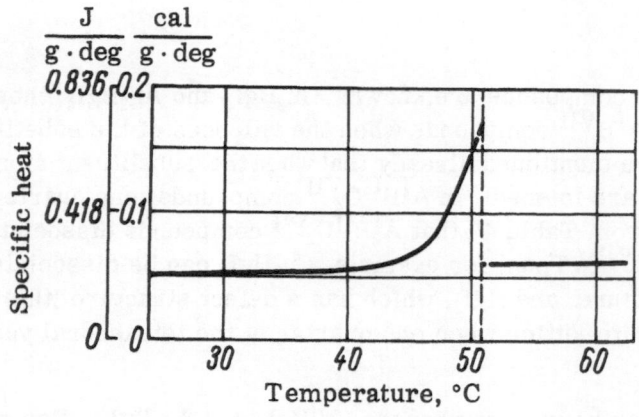

Fig. 48. Temperature dependence of the specific
heat of Ag_2HgI_4.

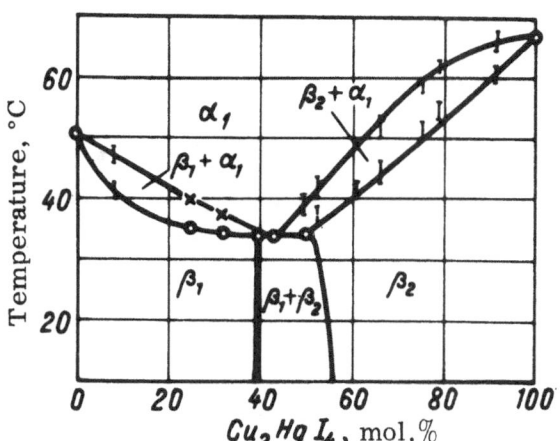

Fig. 49. Phase diagram of the $Ag_2HgI_4-Cu_2HgI_4$ system.

Ag_2HgI_4 have the tetragonal bcc cell with the c parameter twice as large as that found by Ketelaar. The unit cell of β-Cu_2HgI_4 has the following parameters: $a = 0.608_0 \pm 0.0005$ nm $(6.08_0 \pm 0.005$ kX$)$, $c = 1.221_8 \pm 0.0001$ nm $(12.21_8 \pm 0.001$ kX$)$, $c/a = 2.01_1$, while the unit cell of β-Ag_2HgI_4 has the parameters: $a = 0.630_4 \pm 0.0005$ nm $(6.30_4 \pm 0.005$ kX$)$, $c = 1.260_8 \pm 0.0009$ nm $(12.60_8 \pm 0.009$ kX$)$, $c/a = 2.00_0$. Determination of the cell dimensions and of the pycnometric density shows that the unit cell contains two formula units. It has been established that β-Cu_2HgI_4 has the space group D_{2d}^{11} and β-Ag_2HgI_4 has the space group S_4^2 (cf. Fig. 37). Thus, the thiogallate structure is not limited to ternary chalcogenides and can be observed in compounds of other types.

Ag_2HgI_4 exhibits ionic conductivity with a small temperature coefficient. The relatively high ionic conductivity of this compound is due to the low value of the energy for the displacement of ions from their sites. Moreover, the high conductivity is favored by the presence of vacant metal sites in the crystal structure [278, 279]. Figure 47 shows the temperature dependence of the electrical conductivity of this compound.

It is evident from Fig. 47 that at 40°C the electrical conductivity begins to increase very rapidly. At 50.7°C, a structure transition is observed. The electrical conductivity of α-Ag_2HgI_4 is described satisfactorily by the formula $\sigma = A \exp(-B/T)$, where $A = 400$ and $B = 4300$ (in the temperature range 50-100°C). The electrical conductivity of β-Ag_2HgI_4 is more complex.

An investigation of the temperature dependence of the specific heat of Ag_2HgI_4 has indicated a discontinuity in the structure transition region [280]. The temperature dependence of the specific heat (Fig. 48) shows, beginning from 37°C, an anomalous rise: the specific heat increases by a factor of three near the phase transition point. In Ketelaar's opinion, the heterogeneous transition at the discontinuity of the specific heat curve is preceded by transitions which do not disturb the homogeneity but produce a gradual disordering of the structure and which cause the observed increase in the specific heat.

An interesting property of $A_2^I B^{II} C_4^{VII}$ compounds is their thermochromic effect: on transition from the α to the β modification Cu_2HgI_4 changes its color from brown to red, while Ag_2HgI_4 changes from red to yellow. This change of color is gradual. Thus, for example, the change from yellow to orange in Ag_2HgI_4 begins at 40°C and continues up to 50.7°C; above this temperature the red color of the α modification is observed.

Solid solutions of Cu_2HgI_4 and Ag_2HgI_4 have been investigated [281] for the purpose of determining the dependences of their structure, their phase transition point and their color on the composition. The solid solutions were prepared by precipitation of the required substance. It was found that at room temperature the $Cu_2HgI_4-Ag_2HgI_4$ system forms solid solutions with a miscibility gap between 39 and 54 mol.% Cu_2HgI_4. The phase diagram of the $Cu_2HgI_4-Ag_2HgI_4$ system is shown in Fig. 49.

X-ray diffraction analysis of compositions close to the eutectoid point, at temperatures above the phase transition temperature, has indicated a single cubic phase with the lattice parameter $a = 0.6250$ nm (6.250 Å). This parameter is close to the average value of the lattice parameters of the high-temperature modifications of the two original compounds. In other words, the Vegard law is satisfied approximately. All the β phases crystallize in lattices with a pseudocubic tetragonal symmetry.

In the range of compositions from 25 to 50 mol.% Cu_2HgI_4 solid solutions exhibit a rapid change of color from orange to red at temperatures lower than those necessary for the observation of change of color in the original compounds. Some of the other compositions show two more or less gradual color transitions. All these transitions take place at temperatures which are reproducible to within ± 1 deg and when samples are cooled, the color always changes in the direction of shorter wavelengths.

The reflection spectra have been measured for these solid solutions. An increase of the concentration of Cu_2HgI_4 has been found to shift the reflection edge from 510 to 610 nm in the β phase and from 560 to 680 nm in the α phase. The thermochromic effect in $A_2^I B^{II} C_4^{VII}$ compounds and in their solid solutions may possibly be used to prepare temperature indicators.

Hypothetical Ternary Compounds

The criteria for the formation of ternary compounds, given in Chapter I, should be regarded as necessary but insufficient for the formation of ternary compounds with the tetrahedral coordination of atoms, since they take into account only the number of electrons and not the parameters of the atomic cores. Therefore, many predicted compounds have not yet been prepared.

This applies, in particular, to $A^{II} B^{IV} C_2^V$ compounds, of which nitrides, phosphides, and arsenides are known but not antimonides. The attempts of many investigators to prepare $A^{II} B^{IV} C_2^V$ compounds in which the group V element is antimony have not been successful [10, 197, 249, 251, 282]. The causes of the instability of such antimonides have not yet been satisfactorily explained. According to Folberth [13], in the presence of heavy elements there is a competing tendency to form metallic phases instead of compounds, which can formally be crystallized in the tetrahedral system. As the number of components is increased, this tendency should naturally increase so that the electronic analog of InSb, which is the ternary compound $CdSnSb_2$, does not exist in the free state but forms a mixture of two phases: CdSb and SnSb.

Goodman [10, 197] pointed out the lower melting points of $A^{II} B^{IV} C_2^V$ compounds, compared with $A^{III} B^V$ compounds, and of $A^I B^{III} C_2^{VI}$ compounds compared with $A^{II} B^{VI}$ materials, and suggested that the absence of antimony compounds is due to the effect of the low melting point of the corresponding binary antimonides (for example, InSb). Goodman explains the absence of compounds containing lead and mercury by the same reasoning.

Attempts to approach the problem of the stability of ternary compounds more rigorously have been few in number and limited to rough calculations. The polarization of the valence electrons in $A^{II} B^{IV} C_2^V$ compounds is directed toward the C^V atoms but the degree of polarization of $A^{II} - C^V$ and $B^{IV} - C^V$ bonds is different. The chalcopyrite or zinc-blende structure is obtained when each atom has, on the average, four valence electrons. Thus, $A^{II} B^{IV} C_2^V$ compounds are formed more easily when the A^{II} atoms have a strong tendency of accepting the necessary two additional electrons. If we take into account the B^{IV} atoms participating in the formation of these compounds and having the necessary number of electrons, we may conclude that the tendency for the attachment of electrons to the A^{II} atoms should predominate over the corresponding tendency of the B^{IV} atoms. In fact, this does not occur; but we may assume that the formation of a structure with the tetrahedral distribution of atoms becomes easier when the electronegativity of the A^{II} atoms increases relative to the B^{IV} atoms.

TABLE 43. Ratios of Bond Polarizations
of Some Ternary Compounds

Compound	P_{II-V}/P_{IV-V}	Compound	P_{III-VI}/P_{III-IV}
BeSiN$_2$	1.87	Al$_2$CO	1.70
ZnSnP$_2$	2.48		
CdSnP$_2$	2.56	In$_2$GeTe	1.92
ZnSnAs$_2$	2.62	Al$_2$GeTe	1.94
CdSnAs$_2$	2.71		
ZnGeP$_2$	3.16	In$_2$PbTe	2.04
ZnSnSb$_2$	3.18	In$_2$GeSe	2.09
CdSiP$_2$	3.28		
CdGeP$_2$	3.28	Al$_2$GeSe	2.13
CdSnSb$_2$	3.32		
ZnSiAs$_2$	3.43	Tl$_2$PbTe	2.21
ZnGeAs$_2$	3.43		
CdSiAs$_2$	3.56	Ga$_2$GeSe	2.38
CdGeAs$_2$	3.56	In$_2$SnTe	2.45
ZnSiSb$_2$	4.86		
ZnGeSb$_2$	4.98	Al$_2$SnTe	2.50
CdSiSb$_2$	5.05		
CdGeSb$_2$	5.19	In$_2$SnSe	2.68

The value of the polarization can be estimated, in arbitrary units, using the formula [200, 283]:

$$I = \frac{E_{ion}}{E_{cov}}(Z_1 + Z_2)^{3/2},$$

where E_{ion} is the additional ionic energy suggested by Pauling; E_{cov} is the energy of covalent binding; and Z_1 and Z_2 are the atomic numbers of the components.

The additional ionic energy, due to the polarization of bonds, is given by the formula

$$E_{ion} = 92.24 \, (x_1 - x_2)^2,$$

where x_1 and x_2 are the electronegativities of the corresponding elements.

The polarity of the valence bonds can be estimated also using values of the electron affinity, which is the universal energy characteristic of the atomic core [177]:

$$P_{AB} = \rho \, \frac{E_B - E_A}{E_B + E_A} \, ,$$

where P_{AB} is the polarity of the valence bond; ρ is the multiplicity of bonds; and E_B and E_A are the electron affinities of the corresponding atoms.

Approximate values of the electron affinity can be calculated from the formula

$$E = \frac{N}{a + e} \left(\frac{e + 2}{2} \right)^{1/e + 2},$$

where N is the electron affinity constant; a is the number of atoms coordinated around a given core; and e is the number of free electrons of that core [177].

In the tetrahedral coordination $a = 4$ and, when the sp^3 hybrid functions are formed by atoms having on the average four valence electrons, we find that $e = 0$; the above formulas can be then used to calculate approximately the values of the electron affinity. Using the electron affinity, we can estimate the polarization of the II−V and IV−V bonds in $A^{II}B^{IV}C_2^{V}$ compounds.

The results of a calculation of the ratios of the polarizations of the II−V and IV−V bonds are given in Table 43. The dashed line of the left-hand side of Table 43 divides existing and hypothetical compounds. The only exception is $CdSnSb_2$ which had not yet been prepared.

It follows from Table 43 that the transition from phosphides and arsenides to antimonides, with the exception of compounds containing tin, is accompanied by a considerable increase in the ratio of the polarizations P_{A-C}/P_{B-C}, which indicates that the positions of the electron density maxima become less favorable for the formation of tetrahedral structures. Of all the ternary $A^{II}B^{IV}C_2^{V}$ antimonides, the most likely to exist are $CdSnSb_2$ and $ZnSnSb_2$ (the latter compound has been recently prepared by the authors of the present monograph). The existence of ternary antimonides containing germanium and silicon is less likely.

The polarization of the valence bonds affects also the type of structure. As reported in Chapter III, a transition from nitrides, in which these bonds are most strongly polarized, to phosphides produces a transition from the wurtzite to the chalcopyrite structure, while a transition to arsenides gives rise to the zinc-blende-type high-temperature modifications.*

Folberth and Pfister [127] considered the structure of $A^{II}B^{IV}C_2^{V}$ compounds from the point of view of the ratio of the ionic and covalent radii of the A^{II} and B^{IV} elements. They concluded that the ratio $(r_A/r_B)_{ion}$ is always larger than $(r_A/r_B)_{cov}$, which is the consequence of the stronger polarization of the II−V bonds compared with that of the IV−V bonds. Folberth and Pfister assume that the difference between these ratios may be the criterion of the formation of either the chalcopyrite or the zinc-blende structure, since a smaller difference between the ratios of these radii indicates a smaller difference between the bond polarizations and the po-

---

*E. O. Osmanov, Dissertation for Candidate's Degree [in Russian], Silicate Chemistry Institute, Leningrad (1965).

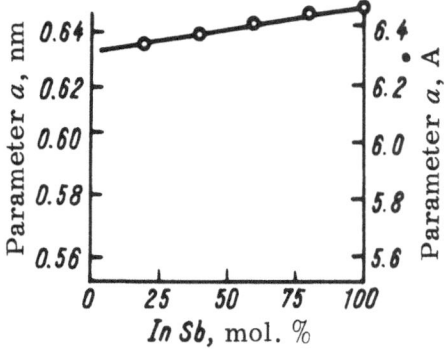

Fig. 50. Dependence of the lattice parameter on the composition in the InSb−ZnSnSb₂ system.

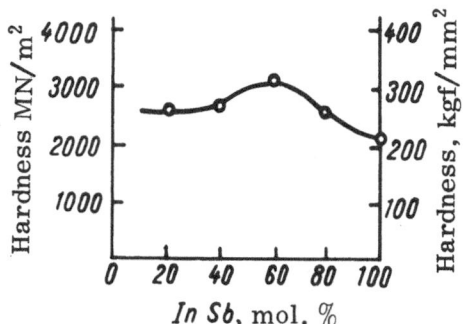

Fig. 51. Dependence of the microhardness on the composition of the InSb−ZnSnSb₂ system.

larization is essential for the appearance of the zinc-blende structure. However, it is unlikely that these considerations can account fully for the stability or instability of ternary compounds.

Palatnik et al. [146] suggested a crystalline geometry criterion for the formation of ternary chemical compounds with the tetrahedral coordination of atoms. These authors took into account the role of the size and attempted to use the well-known calculation of Magnus, carried out for the case of contact between spherical ions. Palatnik et al. [146] assumed this criterion can be used to explain the instability of some ternary compounds. Thus, they concluded that Cu_2SnTe_3 should exist because $r_{Sn^{4+}}/r_{Te^{2-}} = 0.36$, while Cu_2PbTe_3 is hypothetical because $r_{Pb^{4+}}/r_{Te^{2-}} = 0.41$. However, the Magnus criterion applies only to compounds with typically ionic nature of binding and ternary compounds are not of this type. Moreover, Palatnik et al. ignored the fact that the ionic radii of elements of the first group (A^+) are considerably larger (for example, the ionic radius of Cu^+) than the ionic radii of the corresponding elements in the fourth group so that the ratio $r_{A^+}/r_{C^{2-}}$ for $A_2^I B^{IV} C_3^{VI}$ compounds, which are known to exist, does not fit within the limits $0.415 \geq r_c/r_a \geq 0.225$, where r_c is the cation radius and r_a is the anion radius. An experimental confirmation of our view is provided by the fact that the compound Cu_2SnTe_3 does exist, while the compound Ag_2SnTe_3 is hypothetical, although the ratios of their ionic radii do not differ greatly. The inapplicability of the Magnus criterion to ternary semiconducting compounds has already been mentioned in Chapter II.

It is likely that the causes preventing the formation of some ternary compounds are many and that to consider them fully one must take into account, for example, the degree of ionicity and energy of chemical bonds, the covalent and ionic radii of the elements forming a compound, as well as the proportions of atoms belonging to different groups and their valences.

However, the fact that some ternary compounds cannot be prepared in the free form does not mean that solid solutions cannot be formed on the basis of stable compounds having the tetrahedral coordination. In this case, the lattice of the stable compound has a stabilizing influence on the unstable lattice of the ternary compound.

According to Folberth [13], the formation of solid solutions in systems of the $A^{II}B^{IV}C_2^V-A^{III}B^V$ type is due to the fact that, in this case, the polarization of the II−V bonds decreases to the polarization of the III−V bonds, while the polarization of the IV−V bonds increases correspondingly. Thus, the net result is that the polarizations of the three types of bonds assume values which are closer together and this reduces distortions and results in a gain of the energy, which has a favorable effect on the formation of solid solutions. Folberth's ideas are in agreement with many experimental investigations in which solid solutions of the $A^{II}B^{IV}C_2^V-A^{III}B^V$ type have been formed not only on the basis of the existing $A^{II}B^{IV}C_2^V$ compounds but also on the basis of hypothetical compounds. On the other hand, we must mention large discrepancies

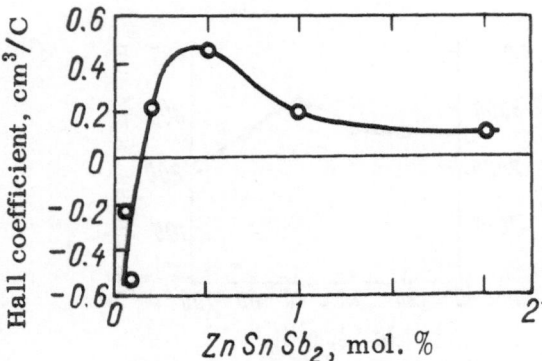

Fig. 52. Dependence of the Hall coefficient on the composition of alloys of the InSb−ZnSnSb$_2$ system.

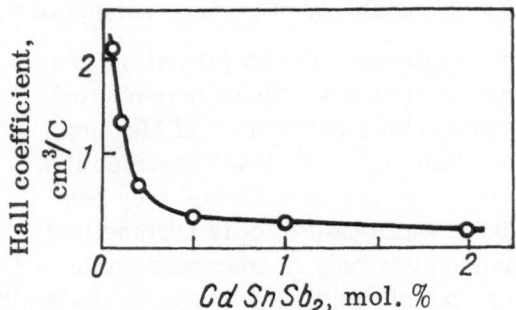

Fig. 53. Dependence of the Hall coefficient of alloys of the InSb−CdSnSb$_2$ system.

between the experimental results obtained by different workers. It has been reported in [252, 318] that zinc-blende solid solutions exist in the system $xZnSnSb_2-(1-x)InSb$ only up to 80% ZnSnSb$_2$. The lattice parameter of this system varies linearly with the composition but the microhardness has a curve with a maximum shifted in the direction of InSb (Figs. 50 and 51).

These solid solutions have been homogenized by quenching, followed by annealing and zone recrystallization in an argon atmosphere. After synthesis, samples of different compositions have been found to exhibit n-type conduction and carrier densities up to 10^{19} cm^{-3}. Woolley and Williams [282] have found a considerably narrower range of existence of solid solutions in the system $xZnSnSb_2-(1-x)InSb$. According to Woolley and Williams, an x-ray diffraction investigation of quenched and annealed samples of this system has indicated the presence of a second phase in the Debye patterns of compositions corresponding to 15-20% ZnSnSb$_2$. Changes in the lattice parameter, found by Woolley and Williams in the range where the second phase has been observed, are not in agreement with those reported in [318]. The dependence of the Hall coefficient on the composition of InSb−ZnSnSb$_2$ alloys is given in Fig. 52.

According to Woolley and Williams, the high carrier density in such alloys is due to the fact that, on the basis of thermodynamic data, we should expect B^{IV} to be partly dissolved in the C^V lattice, resulting in an excess of A^{II} in the A^{III} lattice and a high number of acceptors, which gives rise to a high hole density. These observations are in conflict with the data reported in [252] but are in good agreement with the results given in [254], where a wide range of solid solutions, with p-type conduction and a high carrier density (up to $2.2 \cdot 10^{19}$ cm^{-3}), has been reported for the system $xCdSnAs_2-(1-x)InAs$.

The values of the solubility of CdSnSb$_2$ in InSb, reported in [250, 282] differ considerably from the results given in [249]. According to [282], annealing of samples of 0-50% CdSnSb$_2$ composition at 400°C for 12 weeks does not alter the lattice parameter and, beginning from 10% CdSnSb$_2$, the Debye diffraction patterns show lines representing a second phase, which is also in agreement with the results of a microstructure analysis. All these samples have been found to exhibit p-type conduction; the carrier density is $4 \cdot 10^{19}$ cm^{-3} for 2% CdSnSb$_2$.

Figure 53 shows the dependence of the Hall coefficient on the composition of InSb−CdSnSb$_2$ alloys.

The range of existence of homogeneous phases in this system of alloys is wide and, if we use the formula $xCdSnSb_2-(1-x)(2InSb)$, we find that we can obtain alloys with a small number of inclusions of a second phase up to 50 mol.% CdSnSb$_2$, which corresponds to 33 mol.%, in the formula $xCdSnSb_2-(1-x)InSb$ [249]. In our opinion, such a discrepancy between the reported results is due to different techniques used in the homogenization of these alloys. We must bear in mind that the diffusion processes during the homogenization of semiconducting solid solutions are extremely difficult because of the rigidity and directionality of the covalent bonds. Ignorance of this factor has frequently led to wrong conclusions about the absence of solid solutions. As an example, we can cite the history of the investigations of InSb−AlSb and InSb−GaSb systems, where an insufficiently careful approach to the homogenization gave rise initially to the wrong conclusion that the phase diagrams of these systems are degenerate eutectics [285]. Later investigations have shown that this conclusion is wrong and that solid solutions exist in these systems throughout the whole range of concentrations [286, 319].*

We have established that in the case of annealing below the solidus temperature of nonequilibrium alloys, the homogenization processes take place extremely slowly. The annealing temperatures must exceed solidus temperature of the corresponding nonequilibrium alloys and should be as high as possible. The tendency to retain the short-range order above the nonequilibrium solidus temperature favors the formation of single-phase alloys with the zinc-blende structure.

Moreover, the rate of cooling of the melt during quenching is very important. In some cases, we have placed substances in graphite crucibles and have broken an ampoule during quenching so that the cooling liquid, penetrating through pores in the graphite crucible, has come into direct contact with the melt. We have found that this has reduced the grain size in alloys by an order of magnitude, compared with the grain size in the case of quenching in an unbroken ampoule. This reduction of the grain size by one order of magnitude has shortened very considerably the duration of annealing required after this treatment but even then such duration has sometimes amounted to several thousand hours.

Comparison of the results of microstructure and x-ray diffraction analyses, as well as investigations of the microhardness of alloys of the $CdSnSb_2$−InSb system, have led us to the conclusion that the range of existence of solid solutions in this system is considerably wider than that reported by Woolley and Williams [282].

In addition to investigating this system, we have also determined the possibility, in principle, of the formation of solid solutions between InSb and its ternary electronic analogs of other types. Of the ternary two-cation analogs of InSb (by analogs we understand those ternary tetrahedral phases which are formed by elements of one subgroup of the periodic table), only $AgInTe_2$ exists and it forms a wide range of zinc-blende solid solutions with indium antimonide. Other compounds are hypothetical and direct synthesis produces mixtures of binary compounds and substances in the elemental state. It has been found also that a suitable heat treatment produces solid solutions of InSb with the hypothetical compounds $Ag_2Sn_2Sb_3$, Ag_3SbTe_4, and Ag_2SnTe_3. The ranges of existence of these solid solutions have not been determined but physicochemical investigations have definitely indicated the existence of homogeneous phases near InSb.

Thus, many ternary hypothetical compounds can exist in solid solutions with stable semiconducting compounds. However, investigations of alloys containing hypothetical compounds are, at present, only of theoretical interest. The high carrier density in such alloys and the imperfections of the samples obtained suggest little chance of practical applications of semiconducting properties of such solid solutions.

---

*See also V. I. Ivanov-Omskii, Dissertation for Candidate's Degree [in Russian], State Optical Institute, Leningrad (1961).

Single-Cation Ternary Compounds

We have already pointed out that single-cation nondefect ternary compounds with the tetrahedral coordination of atoms can be divided theoretically into five different types of compound. However, the majority of these compounds have not yet been prepared.

Such compounds should have the antichalcopyrite structure. Folberth [13] concludes that the absence of compounds with the antichalcopyrite structure is associated with the polarization of the valence bonds: the polarizations of these bonds should not differ greatly in a stable compound, because only in such a case the valence-electron hybridization is favored by the energy considerations. However, in the presence of elements of the VIIB subgroup, the difference between the bond polarizations may be very considerable.

The published information on $A_2^{III}B^{IV}C^{VI}$ compounds is very limited. A typical compound of this type, consisting of light elements, is aluminum oxycarbide, which can be regarded as an electronic analog of aluminum nitride. Al_2CO crystals have been prepared by a reaction between Al_2O_3 and Al_4C_3 at a temperature of about 2000°C [288]. Aluminum oxycarbide has a hexagonal crystal lattice of the wurtzite type with the following parameters: $a = 0.317$ nm (3.17 Å), $c = 0.506$ nm (5.06 Å), $c/a = 1.60$. The unit cell of Al_2CO has dimensions 2% larger than aluminum nitride and a corresponding larger value of c/a [289, 320]. The density of Al_2CO found from the x-ray diffraction data is 3.09 g/cm^3, while the experimentally determined density ranges from 2.94 to 3.05 g/cm^3. Single crystals up to 3 mm long have been obtained: they are of needle-like shape with a hexagonal cross section.

There is little information on compounds consisting of heavier elements. We have attempted to synthesize some $A_2^{III}B^{IV}C^{VI}$ compounds containing gallium. We have established that the direct fusion of elements does not produce the ternary compound but follows the reaction: $2A^{III} + B^{IV} + C^{VI} = (\frac{1}{3})A_2^{III}C_3^{VI} + (\frac{4}{3})A^{III} + B^{IV}$. The presence of pure gallium and germanium, found in an attempt to synthesize Ga_2GeSe, confirms this reaction.

Nevertheless, Goodman has synthesized unstable Al_2SnTe, Al_2GeTe, and Al_2GeSe, as well as more stable In_2GeSe and In_2GeTe, which have very complex systems of lines in the Debye diffraction patterns [10].

Woolley and Williams [282] have demonstrated that In_2GeTe can crystallize in the zinc-blende structure when it forms solid solutions with InSb. Figure 54 shows the dependence of

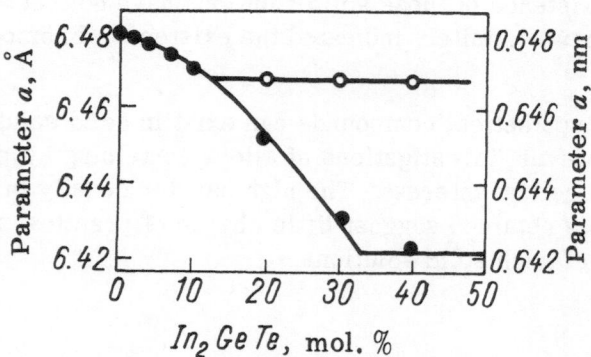

Fig. 54. Dependence of the lattice parameter a on the composition of alloys of the InSb–In_2GeTe system.

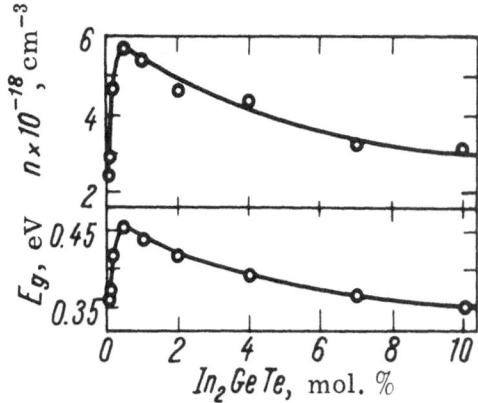

Fig. 55. Dependence of the carrier density and of the forbidden band width on the composition of alloys of the InSb–In$_2$GeTe system.

the lattice parameter a on the composition of annealed and unannealed alloys. It is interesting to note that all alloys have n-type conduction and high carrier densities. Figure 55 shows the dependence of the carrier density and of the forbidden band width, determined from the optical transmission data, on the composition of these alloys.

Similar results have been obtained in investigations of solid solutions in the InSb–In$_2$SnSe and InSb–In$_2$SnTe systems.

Some information on $A_2^{III}B^{IV}C^{VI}$ compounds consisting of heavy elements is given in [320], which reports the existence of the compounds In$_2$PbTe and Tl$_2$PbTe, having a disordered zinc-blende structure. Unfortunately, the existence of these compounds and their structure have not yet been confirmed by other workers. If they are confirmed, the explanations of the instability of ternary compounds on the basis of a competition of the formation of metallic and nonmetallic phases in the presence of heavy elements or on the basis of the lowering of the melting point will require revision. Moreover, if it is found that if the compounds formed by elements in the middle part of the periodic system are either unstable and cannot be prepared by direct synthesis or have complex and unidentified structures, while compounds of lighter elements are stable and have the zinc-blende structure, this will show that the tendencies to form chalcopyrite and antichalcopyrite (zinc blende with two different anions) structures from heavier elements act in opposite directions.

The chalcopyrite or zinc-blende structures, which are due to the formation of tetrahedral sp^3 hybrid bonds, should appear mainly in two-anion ternary compounds only when the polarization of bonds along the anion directions are not too different.

Goryunova [284] pointed out that a transition in a series of binary analog compounds from one crystal structure to another is accompanied by a large change in the electron affinity constants of atoms. Such a correlation between the electron affinity and the crystal structure is more difficult to establish for two anion ternary compounds. When we estimate the polarization of the III–IV and III–VI bonds for all the two-cation compounds described above, using the electron affinities of the corresponding elements, we find that the polarizations are closest for Al$_2$CO. Table 43 lists, on the right-hand side, the ratios of the polarizations P_{III-VI}/P_{III-IV} for $A_2^{III}B^{IV}C^{VI}$ compounds. The wurtzite structure of Al$_2$CO, which has the maximum polarization of the bonds and the minimum ratio of the polarizations, can be predicted from this point of view but the absence of compounds of gallium is difficult to explain. It must be stressed that the approach to the problem of the stability and structure of compounds on the basis of the electron affinity alone is definitely one-sided and primitive.

Relationships Between Physical Properties

A key aspect of solid-state physics and, in particular, in the physics of semiconductors is the preparation of substances with specified properties, i.e., the possibility of predicting the properties of a nonexistent compound using some basic criteria. Such criteria may be, for example, the positions of the components of a compound in the periodic table, the principal quantum number, the atomic weight, etc. [313].

One of the most important characteristics of semiconducting compounds is the forbidden band width. Therefore, many attempts have been made to predict theoretically the forbidden band width on the basis of some concept or another.

Shalyt [293] noted a relationship between the properties of semiconductors and their positions in the periodic table. It can be regarded as established [294, 314, 321] that the forbidden band width depends on the nature of the interatomic interaction, particularly on the ratio of the ionic and metallic components of the binding forces [295, 322].

Sirota [296] found a correlation between the forbidden band width and the energy of the crystal lattice. The higher the lattice energy, the wider is the forbidden band. Manca [297] presented an expression for the dependence of the forbidden band width on the binding energy:

$$\Delta E = a\,(E_s - b),\tag{1}$$

where E_s is the energy of a single bond; and a and b are constants which are fixed for a given type of substances.

In our opinion, the concept of the electronegativity is very useful in estimates of the forbidden band widths of semiconductors. An interesting investigation of this point has been published by Ormont [321], in which the following formulas are suggested for a group of diamond-like semiconductors of the A^{IV}, $A^{II}B^{VI}$, and $A^{III}B^{V}$ types:

$$\Delta E = \left(\frac{v_B}{v_A}\right)^n \cdot [C - M + P]\sum_{hkl}\tag{2}$$

and

$$\Delta E = \left(\frac{v_B}{v_A}\right)^n [c - m + p]\,\Omega,\tag{3}$$

where v_B/v_A is the ratio of the effective charges of atoms forming a compound; M and m are quantities which depend on the total atomic numbers of these atoms; P and p are functions of the electronegativity differences; Σ_{hkl} is the specific surface energy; Ω is the atomization energy; and n, C, and c are constants.

The values of the forbidden band width, calculated using Eqs. (1), (2), and (3), are in satisfactory agreement with the experimental results for some compounds but not for all.

The problem of a relationship between the forbidden band width and the polarization of atoms has been considered in [308].

Presnov [323] used the following expression to calculate the forbidden band width:

$$\Delta E = A\left(\frac{z^*}{r \cdot n} - B\right),\tag{4}$$

where z^* is the effective charge of a nucleus, calculated using Slater's method; r is the interatomic distance; n is the quantum number of the valence electrons; and A and B are constants.

As pointed out in [298, 299], the ratio z^*/r should be regarded as the electronegativity and therefore Eq. (4) is also based on the use of this parameter.

TABLE 44. Forbidden Band Widths
of Some Ternary Compounds [86]

Compound	ΔE, eV		Compound	ΔE, eV	
	calc.	exper.		calc,	exper.
CuAlS₂	2,6	2,5	CuInTe₂	0,1	0,9
CuInS₂	1,3	1,2	AgInTe₂	0,8	1,0
AgInS₂	1,6	1,9	ZnGeP₂	1,9	2,0
CuGaSe₂	1,4	1,6	CdGeP₂	1,4	1,8
CuInSe₂	0,4	0,9	ZnSiAs₂	1,8	2,1
AgInSe₂	0,9	0,7	ZnGeAs₂	0,8	0,8
CuGaTe₂	0,3	1,0			

TABLE 45. Electronegativities of Some Elements [86]

Element	X	Element	X	Element	X
Cu	2,1	In	1,76	P	2,1
Ag	1,85	Tl	1,8	As	1,98
Zn	1,6	Si	1,7	Sb	1,8
Cd	1,7	Ge	1,9	Bi	1,8
Hg	1,8	Sn	1,85	O	3,5
Fe	2,0	Pb	1,8	S	2,6
Al	1,5	Sr	1,3	Se	2,4
Ga	1,7	N	3,0	Te	2,2

Calculations of the forbidden band widths of ternary compounds were carried out first by Batsanov [86], who also used the electronegativity concept. He proposed the following expression for the forbidden band width:

$$\Delta E_{AB} = \Delta E_A + \Delta E_B + a\Delta X_{AB} - b\bar{n},$$

(5)

where ΔE_A and ΔE_B are the values of the forbidden band widths of the components of a compound; ΔX_{AB} is the difference between the electronegativities of the components; $n = \Sigma c_i n_i / \Sigma c_i$ is the average principal quantum number; a and b are constants; n_i is the principal quantum number of the i-th component; and c_i is the number of atoms of a given element in a compound.

In the case of complex (two-cation) compounds, the electronegativity of the cation component is defined as the arithmetic mean of the electronegativities of the atoms forming the cation part of the lattice, taking into account their concentrations:

$$\bar{X}_A = \frac{\Sigma c_i X_i}{\Sigma c_i}.$$

The results of calculations of the forbidden band width using Eq. (5) are given in Table 44 together with the experimental values, while the values of the electronegativities used in these calculations are listed in Table 45.

It is evident from Table 44 that a fully satisfactory agreement between the calculated and experimental values of the forbidden band widths is obtained for the majority of ternary two-cation compounds listed in that table.

In view of this, it is worth considering in detail the role of the electronegativity as a parameter suitable for estimating semiconducting and physicochemical characteristics of substances. An analysis of a correlation between the forbidden band widths and the electronegativity [300] shows that there is no regular relationship between these parameters. This may lead us to conclude that the concept of the electronegativity is completely useless. In our opinion, this concept should neither be regarded as the most important characteristic of a substance nor should we reject its usefulness in those cases when it can be employed with advantage in calculations.

The dependence of the forbidden band width on some other solid-state parameters is discussed in several papers [165, 322, 324, 325]. Belotskii [325] has demonstrated the existence of a linear relationship between the forbidden band width, the melting point, and the total atomic number of some compounds with the zinc-blende structure. The problem of a correlation between the forbidden band width and the ratio of the melting point T_{mp} to the molecular weight M is considered by Berger [324]. Berger [324] is of the opinion that the ratio T_{mp}/M is an implicit function of the crystal lattice energy. This ratio is usually one of the first parameters which can be determined after the preparation of a new substance and therefore it is convenient to use it in preliminary estimates of properties.

Berger [324] points out a linear relationship between ΔE and T_{mp}/M for elemental semiconductors of group IV and for some compounds of the $A^{II}B^{VI}$ type. Comparison of the values of the forbidden band width with the ratio of the melting point and the molecular weight for $A^{III}B^V$ semiconductors indicates a dependence of the type

$$\Delta E \approx (T_{mp}/M - A)^{1/2},$$

where A is a constant, which is approximately equal to 3 for compounds of this type.

The forbidden band width of the majority of known ternary compounds, described in the present monograph, decreases with increasing molecular weight in groups of compounds of a given type and this is accompanied by a reduction of the melting point. The problem of a correlation between properties of ternary compounds is discussed in [165], where the properties of $A_2^I B^{IV} C_3^{VI}$ are considered.

A correlation between the carrier mobility and other parameters of diamond-like semiconducting compounds is discussed in [322, 325]. Goodman [322] noted a relationship between the carrier mobility and the parameter $d^3 E_{ion}$, where d is the lattice constant and E_{ion} is the energy of the ionic component of the interatomic forces. However, the values of the mobilities of some substances do not fit Goodman's relationship. This problem was discussed also by Belotskii [325], who pointed out the existence of a linear relationship between the logarithm of the mobility and the total atomic number of some $A^{III}B^V$ compounds, having a common component of the fifth group, and $A^{II}B^{VI}$ compounds, having a common component of the sixth group.

In all the cited investigations, the dependences of the forbidden band width and of the carrier mobility on various parameters have not been obtained from any theoretical models but have been usually deduced empirically. In our opinion, we should expect the existence of a definite relationship between the forbidden band width and the interatomic interaction energy (taking account of the different forms of the interatomic binding and crystal structure), because these parameters govern the one-dimensional dependence of the potential on the wave vector as well as the shape of the Brillouin zones.

MacDonald and Roy [304] used the one-dimensional Born model to show that the dissociation energy of the crystal lattice, calculated per one pair of atoms, can be expressed in terms of the thermal expansion coefficient:

$$U_0 = \frac{k}{2} \cdot \frac{m+n+3}{m \cdot n} \cdot \frac{1}{\alpha} \, ,$$

where k is the Boltzmann constant; α is the thermal expansion coefficient; and m and n are coefficients which occur in the Born formula:

$$U = -\frac{A}{r^m} + \frac{B}{r^n} \, .$$

This discussion of the relationship between the forbidden band width and the lattice energy, as well as the results reported by MacDonald and Roy [304], suggest that there should be a definite relationship between the principal electrical parameters of a semiconductor and its thermal expansion coefficient.

Zhuze [4] and Kontorova [303] demonstrated that the thermal conductivity of the crystal lattice $\varkappa$ is also related to the lattice energy. At temperatures exceeding the Debye temperature, the following expression is given in [303]

$$\varkappa \propto \frac{(mn)^{3/2}}{(m+n+3)^2} \cdot \frac{U_0^{3/2}}{\sqrt{M} \, r_0^2 kT} \, ,$$

where M is the average mass of atoms; and r_0 is the equilibrium distance between atoms (in the one-dimensional model this distance is the lattice constant).

A correlation between the thermal conductivity and the thermal expansion coefficient is given by the expression [303]

$$\frac{1}{\varkappa} \propto r_0^2 \sqrt{mn} \cdot \sqrt{U_0 M} \, T \alpha^2.$$

Calculations carried out by Berger [305] show that the lattice thermal conductivity, at temperatures higher than the Debye temperature, should have the following temperature dependence:

$$\varkappa = B \sqrt{\frac{T_{mp}}{M}} \cdot \frac{1}{T} \, , \tag{6}$$

where B is a numerical factor.

In spite of many simplifying assumptions, Eq. (6) is in satisfactory agreement with the experimental data.

A relationship between the thermal conductivity and thermal expansion of many diamond-like substances has been considered also in [118, 145].* It has been found that the following relationship

$$\frac{1}{\varkappa} \propto \alpha^2$$

is valid for diamond-like substances in a wide range of temperatures.

---

*See also L. I. Berger, Dissertation for Candidate's Degree [in Russian], M. I. Kalinin Institute for Nonferrous Metals and Gold, Moscow (1958).

It is shown in [160] that some $A_2^{I}B^{IV}C_3^{VI}$ compounds obey the following relationships at room temperature:

$$\varkappa \propto (T_{mp}/M)^{1/2},$$ (7)

$$\alpha \propto (T_{mp}/M)^{-1/4},$$ (8)

$$E \propto \rho \cdot T_{mp}/M,$$ (9)

where E is Young's modulus and ρ is the density. The possibility of existence of relationships given by Eqs. (7)-(9) has been demonstrated in [305].

Neshpor [306] has suggested an expression, which is of practical importance, relating Young's modulus, the number of atoms in the unit cell, the specific heat c, the Gruneisen constant γ, the molecular weight M, and the thermal expansion coefficient α:

$$E = \frac{42.6 Z c \gamma}{M \alpha}.$$

The existence of a relationship between the hardness and the thermal conductivity of the crystal lattice has been demonstrated for a large group of ternary $A^{I}B^{III}C_2^{VI}$ compounds by Zhuze and Kontorova [134]. Lozinskii and Fedotov [307] have demonstrated the existence of a definite relationship between the hardness and Young's modulus at high temperatures $(T \gg \Theta)$. Kontorova [311] has pointed out a close relationship between the mechanical and thermal properties of crystals. The same conclusions were reached by Ormont [308-310]. The existence of a correlation between the thermal expansion and the permittivity is pointed out in Panchenko's paper [312].

This brief review of investigations concerned with relationships between physical properties shows that a close correlation exists between many electrical, thermal, mechanical, and other properties of crystals. However, the majority of the published relationships have been derived using many simplifying assumptions and they cannot be used for quantitative estimates. It follows that one of the most urgent problems facing investigators working in solid-state and semiconductor physics and chemistry is the accumulation of very extensive experimental data which could be used as a basis for the development of a reasonable theoretical model, supported by experimental evidence, and suitable for the a priori estimates of properties of particular substances.

LITERATURE CITED

1. L. Pauling, Nature of the Chemical Bond, 3rd ed., Cornell Univ. Press, New York (1960).
2. M. G. Veselov, Elementary Quantum Theory of Atoms and Molecules [in Russian], Fizmatgiz (1962).
3. E. Parthé, Crystal Chemistry of Tetrahedral Structures, Gordon and Breach, New York (1964).
4. V. P. Zhuze, Dokl. Akad. Nauk SSSR, 99:711 (1954).
5. E. Mooser and W. B. Pearson, J. Chem. Phys., 26:893 (1957).
6. G. E. Kimball, J. Chem. Phys., 8:188 (1940).
7. D. F. G. Morris and L. H. Ahrens, J. Inorg. Nucl. Chem., 3:263 (1956).
8. H. Krebs, Z. Anorg. Allgem. Chem., 278:82 (1955).
9. E. Mooser and W. B. Pearson, J. Electronics, 1:629 (1956).
10. C. H. L. Goodman, J. Phys. Chem. Solids, 6(4):305 (1958).
11. W. B. Pearson, in: Semiconducting Materials [Russian translation], Izd. AN SSSR, Moscow (1960), p. 249.
12. N. A. Goryunova, Chemistry of Diamond-like Semiconductors, Chapman and Hall London (1965).
13. O. G. Folberth, Z. Naturforsch., 14a:94 (1959).
14. H. Krebs, Acta Cryst., 9:95 (1956).
15. J. C. Eisenstein, J. Chem. Phys., 25:1242 (1956).
16. H. Grimm and A. Sommerfeld, Z. Physik, 36:36 (1926).
17. J. E. Spice, Chemical Binding and Structure, Pergamon Press, Oxford (1964).
18. N. A. Goryunova, Vestnik LGU, 10:112 (1961).
19. G. B. Bokii, Crystal Chemistry [in Russian], Izd. MGU (1961).
20. L. S. Palatnik and E. I. Rogacheva, Izv. Akad. Nauk, Neorgan. Mat., 2:993 (1966).
21. A. J. Strauss and A. J. Rosenberg, J. Phys. Chem. Solids, 17:278 (1961).
22. N. A. Goryunova and E. Parthé, Materials Science and Engineering, USA (1967).
23. O. G. Folberth and H. Pfister, Acta Cryst., 13:199 (1960).
24. L. S. Palatnik, V. M. Koshkin, and L. P. Gal'chinetskii, Fiz. Tverd. Tela, 4:2365 (1962).
25. L. S. Palatnik, Yu. F. Komnik, and V. M. Koshkin, Kristallografiya, 7:563 (1962).
26. L. Pauling and M. L. Huggins, Z. Krist., 87:205 (1934).
27. A. F. Wells, Structural Inorganic Chemistry, 3rd ed., Clarendon Press, Oxford (1962).
28. W. Gordy and W. S. O. Thomas, J. Chem. Phys., 24:439 (1956).
29. R. Sandrock and J. Treusch, Z. Naturforsch., 19a:844 (1964).
30. Strukturberichte, 2:48 (1937).
31. H. Hahn, G. Frank, W. Klingler, A. Meyer, and A. D. Störger, Z. Anorg. Allgem. Chem., 271:153 (1953).
32. B. F. Ormont, Structures of Inorganic Substances [in Russian], Gostekhizdat, Moscow (1950).
33. A. Rabenau and P. Eckerlin, Naturwiss., 46:106 (1959).

34. F. M. Gashimzade, Fiz. Tverd. Tela, 5:1199 (1963).

35. L. Pauling and L. Brocway, Z. Krist., 82:188 (1932).

36. A. A. Vaipolin, F. M. Gashimzade, N. A. Goryunova, F. P. Kesamanly, D. N. Nasledov, É. O. Osmanov, and Yu. V. Rud', Izv. Akad. Nauk SSSR, Ser. Fiz., 28:1085 (1964).

37. N. P. Luzhnaya, G. F. Nikol'skaya, and I. S. Kovaleva, Izv. Akad. Nauk,Neorgan. Mat., 2:128 (1966).

38. H. Borchers and R. G. Maier, Metall., 17:775 (1963).

39. D. R. Mason and D. F. O'Kane, Proc. Fifth Intern. Conf. on Physics of Semiconductors, Prague, 1960, publ. by Academic Press, New York (1961), p. 1026.

40. J. van den Boomgaard and K. Schol, Philips Res. Rep., 12:127 (1957).

41. O. G. Folberth, Halbleiterprobleme, Vol. 5, Vieweg, Braunschweig (1960), p.40.

42. A. A. Vaipolin and N. M. Korshak, in: Physics (Proc. Twenty-Third Sci. Conf. at Leningrad Structural Engineering Institute) [in Russian], Leningrad (1965), p. 47.

43. K. Schubert, Kristallstrukturen zweikomponentiger Phasen, Berlin (1964).

44. L. Ekstrom and L. R. Weisberg, J. Electrochem. Soc., 109:321 (1962).

45. E. A. Vol, Structure and Properties of Binary Metallic Systems [in Russian], Vol. 1 (1959), Vol. 2 (1962), Fizmatgiz, Moscow.

46. J. Rivet, J. Flahaut, and P. Laruelle, Compt. Rend., 257:161 (1963).

47. L. I. Berger and R. Annamamedov, Izv. Akad. Nauk Turkm.SSR, Ser. Fiz.-Tekhn., Matem. i Geol. Nauk, 2:129 (1965).

48. M. Hansen and K. Anderko, Constitution of Binary Alloys, McGraw-Hill, New York (1958).

49. N. A. Goryunova and V. I. Sokolova, Izv. Mold. Filiala Akad. Nauk SSSR, No. 3(69), p. 31 (1960).

50. E. M. Conwell, Proc. IRE, 46:1281 (1958).

51. F. A. Kröger and D. Nobel, J. Electronics, 1:190 (1955).

52. J. R. Van Wazer, Phosphorus and Its Compounds, 2 vols.,Interscience, New York (1958, 1961).

53. W. G. Spitzer, M. Gershenzon, C. J. Frosch, and D. F. Gibbs, J. Phys. Chem. Solids, 11:339 (1959).

54. Handbook of Chemistry and Physics, 41:2335 (1959-1960).

55. R. F. Mekhtiev, É. O. Osmanov, and Yu. V. Rud', Pribory i Tekh. Eksperim., No. 2, p. 179 (1964).

56. Yu. V. Shmartsev, Yu. A. Valov, and A. S. Borshchevskii, Refractory Semiconductor Materials, Consultants Bureau, New York (1966).

57. Y. Mita, J. Phys. Soc. Japan, 17:784 (1962).

58. G. Wolff, P. H. Keck, and J. D. Broder, Bull. Am. Phys. Soc., 29:16 (1954).

59. T. C. Harman, J. I. Genco, W. P. Allred, and H. L. Goering, J. Electrochem. Soc., 105:731 (1958).

60. Y. Mita, J. Phys. Soc. Japan, 16:1484 (1961).

61. Ya. A. Ugai, Introduction to Chemistry of Semiconductors [in Russian], Izd. "Vysshaya shkola," Moscow (1965).

62. J. W. Faust, Jr. and H. F. John, J. Phys. Chem. Solids, 25:1407 (1964).

63. B. V. Baranov, V. S. Grigor'eva, L. V. Kradinova, and V. D. Prochukhan, in: Physics (Proc. Twenty-Third Sci. Conf. at Leningrad Structural Engineering Institute) [in Russian], Leningrad (1965), p. 48.

64. I. N. Belyaev, Growth of Crystals, Vol. 3, Consultants Bureau, New York (1962), p. 316.

65. J. W. Nielsen and E. F. Dearborn, J. Phys. Chem. Solids, 5:202 (1958).

66. H. Schäfer, Chemische Transportreaktionen: der Transport anorganischer Stöffe über die Gasphase und seine Anwendungen, Verlag Chemie, Weinheim/Bergstr. (1962).

67. R. F. Lever, J. Chem. Phys., 37:1174 (1962).

68. R. Nitsche, H. U. Bölsterli, and M. Lichtensteiger, J. Phys. Chem. Solids, 21:199 (1961).

69. R. R. Möst and B. R. Shapp, J. Electrochem. Soc., 109:1061 (1962).
70. R. Hoppe, Angew. Chem., 71:1457 (1959).
71. H. Samelson, J. Appl. Phys., 33:1179 (1962).
72. F. Jona, J. Phys. Chem. Solids, 23:1719 (1962).
73. R. Nitsche, J. Phys. Chem. Solids, 17:163 (1960).
74. J. A. Beun, R. Nitsche, and M. Lichtensteiger, Physica, 27:148 (1961).
75. S. Shionoya and Y. Tamoto, J. Phys. Soc. Japan, 19:1142 (1964).
76. V. F. Zhitar', N. A. Goryunova, and S. I. Radautsan, Proc. Conf. on Growth of Semi-conductor Single Crystals [in Russian], Izd. SO Akad. Nauk SSSR, Novosibirsk (1966), p. 172.
77. L. Navias, J. Am. Ceram. Soc., 45:544 (1962).
78. N. A. Goryunova, G. K. Averkieva, and A. A. Vaipolin, in: Physics (Proc. Twenty-Third Sci. Conf. at Leningrad Structural Engineering Institute) [in Russian], Leningrad (1965), p. 52.
79. L. I. Berger and V. M. Petrov, in: Chemical Reagents and Materials, Trudy IREA, No. 30, p. 396 (1967).
80. G. Mandel, J. Chem. Phys., 40:683 (1964).
81. N. A. Goryunova, Yu. A. Valov, and L. B. Zlatkin, in: Physics (Proc. Twenty-Third Sci. Conf. at Leningrad Structural Engineering Institute) [in Russian], Leningrad (1965), p. 18.
82. H. Hahn and C. de Lorent, Z. Anorg. Allgem. Chem., 279:281 (1955).
83. M. Elscherbini and J. Jösef, Proc. Phys. Soc. (London), 51:449 (1939).
84. E. T. Wherry, Am. Mineral.,10:28 (1925).
85. H. F. Mataré, Electrochem. Soc. Meeting, Chicago (1954).
86. S. S. Batsanov, Zh. Strukt. Khim., 5:927 (1964).
87. C. H. L. Goodman and R. W. Douglas, Physica, 20:1107 (1954).
88. I. G. Austin, C. H. L. Goodman, and A. E. Pengelly, Nature, 178:433 (1956).
89. I. G. Austin, C. H. L. Goodman, and A. E. Pengelly, J. Electrochem. Soc., 103:609 (1956).
90. R. W. Douglas and C. H. L. Goodman, GEC J., 21:3 (1954).
91. A. Fruch, Am. Mineral., 35:282 (1950).
92. A. R. Regel', Zh. Tekh. Fiz., 18:1511 (1948).
93. V. M. Glazov, M. S. Mirgalovskaya, and L. A. Petrakova, Izv. Akad. Nauk SSSR, OTN, 10:68 (1957).
94. V. P. Zhuze, V. M. Sergeeva, and E. L. Shtrum, Zh. Tekh. Fiz., 28:2093 (1958).
95. L. I. Berger and A. É. Balanevskaya, Izv. Akad. Nauk SSSR, Neorgan. Mat., 2:1514 (1966).
96. A. É. Balanevskaya, L. I. Berger, and I. A. Dobrushina, Abstracts of Papers presented at Third All-Union Conf. on Problems of Chemical Binding in Semiconductors [in Russian], Izd. "Nauka i tekhnika," Minsk (1965), p. 43.
97. J. W. Davisson and J. Pasternak, Status Report on Thermoelectricity, NRL Memorandum Rep. (1960), p. 1037.
98. A. V. Voitsekhovskii, Abstracts of Papers presented at All-Union Conf. on Semiconducting Compounds [in Russian], Izd. AN SSSR, Moscow (1961), p. 13.
99. A. V. Petrov and E. L. Shtrum, Fiz. Tverd. Tela, 4:1442 (1962).
100. R. Annamamedov, A. É. Balanevskaya, L. I. Berger, I. A. Dobrushina, and I. K. Shchukina, in: Chemical Bonds in Semiconductors and Thermodynamics (ed. by N. N. Sirota), Consultants Bureau, New York (1968), p. 240.
101. V. M. Petrov, A. É. Balanevskaya, F. F. Kharakhorin, and L. I. Berger, Izv. Akad. Nauk SSSR, Neorgan. Mat., 2:1874 (1966).
102. A. É. Balanevskaya, L. I. Berger, A. V. Pechennikov, and V. I. Chechernikov, Izv. Akad. Nauk SSSR, Neorgan. Mat., 1:2163 (1965).

103. S. I. Novikova, Fiz. Tverd. Tela, 7:2683 (1965).

104. S. I. Novikova and P. G. Strelkov, Fiz. Tverd. Tela, 1:1841 (1959).

105. A. V. Petrov and E. L. Shtrum, Abstracts of Papers presented at All-Union Conf. on Semiconducting Compounds [in Russian], Izd. AN SSSR, Moscow (1961), p. 53.

106. S. I. Novikova, Fiz. Tverd. Tela, 3:178 (1961).

107. V. S. Oskotskii, Fiz. Tverd. Tela, 6:1294 (1964).

108. V. P. Chernyavskii, in: Physics (Proc. Twentieth Sci. Conf. at Leningrad Structural Engineering Institute) [in Russian], Leningrad (1962), p. 10.

109. V. P. Zhuze, V. M. Sergeeva, and E. L. Shtrum, Zh. Tekh. Fiz., 28:233 (1958).

110. E. L. Shtrum, in: Problems in Metallurgy and Physics of Semiconductors [in Russian], Izd. AN SSSR, Moscow (1961), p. 24.

111. P. Manca and F. Massazza, J. Appl. Phys., 36:647 (1965).

112. O. V. Losev, Telegraf i Telefon bez Provodov, No. 4 (1922).

113. O. V. Losev, Telegraf i Telefon bez Provodov, No. 15 (1922).

114. B. I. Boltaks and N. N. Tarnovskii, Zh. Tekh. Fiz., 25:402 (1955).

115. H. Schlegel and A. Schüller, Z. Metallkunde, 43:421 (1952).

116. L. S. Palatnik, Yu. F. Komnik, V. M. Koshkin, L. P. Gal'chinetskii, and L. G. Manyukova, Ukr. Fiz. Zh., 9:962 (1964).

117. L. I. Berger and I. A. Dobrushina, in: Chemical Reagents and Materials, Trudy IREA, No. 26, p. 302 (1964).

118. N. N. Sirota and L. I. Berger, Inzh.-Fiz. Zh., 2:102 (1959).

119. E. F. Apple, J. Electrochem. Soc., 105:251 (1958).

120. A. É. Balanevskaya and L. I. Berger, Abstracts of Papers presented at Third All-Union Conf. on Problems of Chemical Binding in Semiconductors [in Russian], Izd. "Nauka i tekhnika," Minsk (1965), p. 44.

121. S. I. Novikova and N. Kh. Abrikosov, Fiz. Tverd. Tela, 5:2138 (1963).

122. S. I. Novikova, Fiz. Tverd. Tela, 2:43 (1960).

123. S. I. Novikova, Fiz. Tverd. Tela, 2:2341 (1960).

124. S. P. Bardeeva, I. S. Lisker, and A. F. Chudnovskii, in: Physics (Proc. Twentieth Sci. Conf. at Leningrad Structural Engineering Institute) [in Russian], Leningrad (1962), p. 34.

125. V. M. Koshkin, Ukr. Fiz. Zh., 9:1038 (1964).

126. A. A. Vaipolin, É. O. Osmanov, and Yu. V. Rud', Fiz. Tverd. Tela, 7:2266 (1965).

127. O. G. Folberth and H. Pfister, Acta Cryst., 14:325 (1961).

128. J. P. Remeika and A. A. Ballman, Appl. Phys. Letters, 5:180 (1964).

129. J. Thery, A. M. Lejus, D. Briancon, and R. Collongues, Bull. Chim. France, p. 973 (1961).

130. A. M. Lejus and R. Collongues, Compt. Rend., 254:2005 (1962).

131. M. Marezio and J. P. Remeika, J. Phys. Chem. Solids, 26:1277 (1965).

132. F. Bertaut and P. Blum, Compt. Rend., 239:429 (1954).

133. J. Thery, D. Briancon, and R. Collongues, Compt. Rend., 252:1475 (1961).

134. V. P. Zhuze and T. A. Kontorova, Zh. Tekh. Fiz., 28:1727 (1958).

135. S. I. Radautsan, R. A. Maslyanko, and M. M. Markus, in: Soviet Research on New Semiconducting Materials (ed. by D. N. Nasledov and N. A. Goryunova), Consultants Bureau, New York (1965), p. 101.

136. L. S. Palatnik and E. K. Belova, Izv. Akad. Nauk SSSR, Neorgan. Mat., 2:820 (1966).

137. L. S. Palatnik and E. I. Rogacheva, Kristallografiya, 11:95 (1966).

138. L. S. Palatnik, L. G. Manyukova, and V. M. Koshkin, Izv. Akad. Nauk SSSR, Neorgan. Mat., 2:350 (1966).

139. L. S. Palatnik, E. K. Belova, L. V. Atroshchenko, and Yu. F. Komnik, Kristallografiya, 10:474 (1965).

140. L. S. Palatnik, Yu. F. Komnik, and E. I. Rogacheva, Ukr. Fiz. Zh., 9:862 (1964).

141. L. S. Palatnik and E. K. Belova, Kristallografiya, 10:950 (1965).

142. L. S. Palatnik and E. I. Rogacheva, Izv. Akad. Nauk SSSR, Neorgan. Mat., 2:700 (1966).

143. R. M. Imamov and Z. G. Pinsker, Kristallografiya, 10:284 (1965).

144. R. M. Imamov and Z. G. Pinsker, Kristallografiya, 9:743 (1964).

145. L. I. Berger and S. I. Radautsan, in: Problems in Metallurgy and Physics of Semi-conductors [in Russian], Izd. AN SSSR, Moscow (1961), p. 129.

146. L. S. Palatnik, Yu. F. Komnik, V. M. Koshkin, and E. K. Belova, Dokl. Akad. Nauk SSSR, 137:68 (1961).

147. V. M. Goldschmidt, Usp. Fiz. Nauk, 9:811 (1929) [Russian translation].

148. H. Hahn, XVII Congress IUPAC, Munich, 1959, p. 157.

149. G. K. Averkieva and A. A. Vaipolin, Abstracts of Papers presented at All-Union Conf. on Semiconducting Compounds [in Russian], Izd. AN SSSR, Moscow (1961), p. 4.

150. L. S. Palatnik, Yu. F. Komnik, E. K. Belova, L. V. Atroshchenko, and E. I. Rogacheva, Abstracts of Papers presented at All-Union Conf. on Semiconducting Compounds [in Russian], Izd. AN SSSR, Moscow (1961), p. 50.

151. L. S. Palatnik, Yu. F. Komnik, and V. M. Koshkin, Abstracts of Papers presented at All-Union Conf. on Semiconducting Compounds [in Russian], Izd. AN SSSR, Moscow (1961), p. 51.

152. N. A. Goryunova, A. V. Voitsekhovskii, and V. D. Prochukhan, Vestnik Leningrad Gos. Univ., 10:156 (1962).

153. N. A. Goryunova and Chiang Ping-Hsi, Abstracts of Papers presented at All-Union Conf. on Semiconducting Compounds [in Russian], Izd. AN SSSR, Moscow (1961), p. 21.

154. L. S. Palatnik, V. M. Koshkin, and Yu. F. Komnik, Kristallografiya, 7:124 (1962).

155. L. S. Palatnik, V. M. Koshkin, L. P. Gal'chinetskii, V. I. Kolesnikov, and Yu. F. Komnik, Fiz. Tverd. Tela, 4:1430 (1962).

156. L. S. Palatnik, Yu. F. Komnik, E. K. Belova, and L. V. Atroshchenko, Ukr. Fiz. Zh., 8:263 (1963).

157. F. F. Kharakhorin and V. M. Petrov, Fiz. Tverd. Tela, 6:2867 (1964).

158. G. K. Averkieva, A. A. Vaipolin, and N. A. Goryunova, in: Soviet Research on New Semiconducting Materials (ed. by D. N. Nasledov and N. A. Goryunova), Consultants Bureau, New York (1965), p. 1.

159. L. S. Palatnik, Yu. F. Komnik, E. K. Belova, and L. V. Atroshchenko, Kristallografiya, 6:960 (1961).

160. L. I. Berger and A. É. Balanevskaya, Fiz. Tverd. Tela, 6:1311 (1964).

161. L. I. Berger, Zavod. Lab., 6:712 (1965).

162. J. Rivet, O. Gorochov, and J. Flahaut, Compt. Rend., 260:178 (1965).

163. A. É. Balanevskaya, L. I. Berger, and V. M. Petrov, Izv. Akad. Nauk SSSR, Neorgan. Mat., 2:810 (1966).

164. A. F. Ioffe, Physics of Semiconductors, Infosearch Ltd., London (1960).

165. L. I. Berger and R. Annamamedov, in: Chemical Reagents and Materials, Trudy IREA, No. 29, p. 249 (1966).

166. N. A. Goryunova, G. K. Averkieva, and Yu. V. Alekseev, Izv. Mold. Filiala Akad. Nauk SSSR, No. 3(69), p. 99 (1960).

167. N. P. Luzhnaya, Abstracts of Papers presented at Fourth All-Union Conf. on Physico-chemical Analysis [in Russian], Izd. AN SSSR (1960), p. 86.

168. W. F. de Jong, Z. Krist., 68:522 (1928).

169. C. Hermann, O. Lohrmann, and H. Philipp, Supplement Vol. II, Strukturberichte, Vol. 2 p. 347 (1928-1932).

170. A. W. Waldo, Am. Mineral., 20:575 (1935).

171. R. W. G. Wyckoff, Crystal Structures, Vol. 1, Interscience, New York (1948).

172. A. F. Wells, Structural Inorganic Chemistry, 2nd ed., Clarendon Press, Oxford (1950).

173.   J. H. Wernick and K. E. Benson, J. Phys. Chem. Solids, 3:157 (1957).

174.   G. Busch and F. Hulliger, Helv. Phys. Acta, 33:657 (1960).

175.   P. C. Newman, J. Phys. Chem. Solids, 24:45 (1963).

176.   N. A. Goryunova, G. M. Orlova, A. V. Danilov, A. V. Abramova, R. L. Plechko, and
       I. I. Kozhina, Vestnik Leningrad Gos. Univ., Ser. Fiziki i Khimii, 22:97 (1961).

177.   B. V. Nekrasov, General Chemistry Course [in Russian], Goskhimizdat (1954).

178.   L. I. Berger and R. Annamamedov, Izv. Akad. Nauk SSSR, Neorgan. Mat., 1:511 (1965).

179.   R. Annamamedov, L. I. Berger, V. M. Petrov, and S. V. Slobodchikov, Izv. Akad. Nauk
       SSSR, Neorgan. Mat., 3:1370 (1967).

180.   J. H. Wernick, S. Geller, and K. E. Benson, J. Phys. Chem. Solids, 4:154 (1958).

181.   J. H. Wernick, S. Geller, and K. E. Benson, J. Phys. Chem. Solids, 7:240 (1958).

182.   R. Wolfe, J. H. Wernick, and S. E. Haszko, J. Appl. Phys., 31:1959 (1960).

183.   L. Pauling and S. Weinbaum, Z. Krist., 88:48 (1934).

184.   D. Lundquist and A. Westgren, Swensk. Kem. Tidskr., 48:41 (1936).

185.   L. Pauling and R. Hultgren, Z. Krist., 84:204 (1933).

186.   N. V. Belov, Structure of Ionic Crystals and Metal Phases [in Russian], Izd. AN SSSR
       (1947), p. 59.

187.   A. G. Alieva and Z. G. Pinsker, Kristallografiya, 6:204 (1961).

188.   V. M. Petrov, R. Annamamedov, F. F. Kharakhorin, and L. I. Berger, Izv. Akad. Nauk
       SSSR, Neorgan. Mat., 2:1053 (1966).

189.   L. I. Berger and R. Annamamedov, Abstracts of Papers presented at Third All-Union
       Conf. on Problems of Chemical Binding in Semiconductors [in Russian], Izd. "Nauka
       i tekhnika," Minsk (1965), p. 45.

190.   H. Fleischmann, O. G. Folberth, and H. Pfister, Z. Naturforsch., 14a:999 (1959).

191.   R. W. Armstrong, J. W. Faust, Jr., and W. A. Tiller, J. Appl. Phys., 31:1954 (1960).

192.   N. A. Goryunova and V. I. Sokolova, Izv. Mold. Filiala Akad. Nauk SSSR, No.3(69), p. 192
       (1960).

193.   V. I. Sokolova, Abstracts of Papers presented at All-Union Conf. on Semiconducting
       Compounds [in Russian], Izd. AN SSSR, Moscow (1961), p. 60.

194.   V. I. Sokolova and V. I. Tsvetkova, in: Soviet Research on New Semiconducting Ma-
       terials (ed. by D. N. Nasledov and N. A. Goryunova), Consultants Bureau, New York
       (1965), p. 111.

195.   N. A. Goryunova, Izv. Akad. Nauk SSSR, Ser. Fiz., 21:120 (1957).

196.   O. G. Folberth and H. Pfister, Proc. Third Conf. on Physics of Semiconductors,
       Garmisch-Partenkirchen, 1956, publ. as Halbleiter und Phosphore, Vieweg, Braun-
       schweig (1958), p. 474.

197.   C. H. L. Goodman, Nature, 179:828 (1957).

198.   H. Pfister, Acta Cryst., 11:221 (1958).

199.   A. Stegherr, F. Wald, and P. Eckerlin, Z. Naturforsch., 16a:130 (1961).

200.   O. G. Folberth, Z. Naturforsch., 13a:856 (1958).

201.   D. B. Gasson, P. J. Holmes, I. C. Jennings, B. R. Marathe, and J. E. Parrott, J. Phys.
       Chem. Solids, 23:1291 (1962).

202.   A. J. Rosenberg and A. J. Strauss, Bull. Am. Phys. Soc., 5:83 (1960).

203.   M. Tanenbaum and H. B. Briggs, Phys. Rev., 91:1561 (1953).

204.   E. O. Kane, J. Phys. Chem. Solids, 1:249 (1957).

205.   R. P. Chasmar and R. Stratton, J. Electronics Control, 7:52 (1959).

206.   N. A. Goryunova, S. Mamaev, and V. D. Prochukhan, Dokl. Akad. Nauk SSSR, 142:623
       (1962).

207.   M. Matyáš and P. Höschl, Czech. J. Phys., 12B:788 (1962).

208.   H. Pfister, Acta Cryst., 16:153 (1963).

209.   N. A. Goryunova, F. P. Kesamanly, D. N. Nasledov, and Yu. V. Rud', Fiz. Tverd. Tela,
       6:113 (1964).

210. F. P. Kesamanly, D. N. Nasledov, and Yu. V. Rud', Fiz. Tverd. Tela, 6:2187 (1964).

211. Chiang Ping-Hsi, I. I. Tychina, É. O. Osmanov, and N. A. Goryunova, in: Physics (Proc. Twenty-First Sci. Conf. at Leningrad Structural Engineering Institute) [in Russian], Leningrad (1963), p. 8.

212. P. Leroux-Hugon, Compt. Rend., 256:118 (1963).

213. P. Leroux-Hugon, Compt. Rend., 256:3991 (1963).

214. R. M. Talley and F. Stern, J. Electronics, 1:186 (1955).

215. T. C. Harman, Bull. Am. Phys. Soc., 5:152 (1960).

216. N. A. Goryunova, S. Mamaev, and V. D. Prochukhan, Abstracts of Papers presented at All-Union Conf. on Semiconducting Compounds [in Russian], Izd. AN SSSR, Moscow (1961), p. 20.

217. F. M. Gashimzade, Izv. Akad. Nauk AzSSR, Ser. Fiz.-Mat. Nauk, 3:72 (1963).

218. R. H. Parmenter, Phys. Rev., 100:573 (1955).

219. É. Rashba and V. Sheka, Fiz. Tverd. Tela, 1:162 (1959).

220. O. V. Emel'yanenko and F. P. Kesamanly, Fiz. Tverd. Tela, 2:1494 (1960).

221. A. A. Vaipolin (Vypolin), F. M. Gashimzade, N. A. Goryunova, F. P. Kesamanly, D. N. Nasledov, and E. O. Osmanov, Izv. Akad. Nauk SSSR, Ser. Fiz., 28:1085 (1964).

222. F. M. Gashimzade, Izv. Akad. Nauk AzSSR, Ser. Fiz.-Mat. Nauk, 3:67 (1963).

223. O. V. Emel'yanenko and N. V. Trishin, Pribory i Tekh. Eksperim., No. 1, p. 98 (1960).

224. A. A. Vaipolin (Vypolin), É. O. Osmanov, and D. N. Tretyakov, XXth Intern. Congress IUPAC (Abbr. Engl. transl. of sci. papers) D45, Moscow (1965).

225. N. A. Goryunova, F. P. Kesamanly, and É. O. Osmanov, Fiz. Tverd. Tela, 5:2031 (1963).

226. F. M. Gashimzade, Fiz. Tverd. Tela, 4:2059 (1962).

227. O. V. Emel'yanenko, F. P. Kesamanly, and D. N. Nasledov, Fiz. Tverd. Tela, 3:1162 (1961).

228. Gashimzade and F. P. Kesamanly, Fiz. Tverd. Tela, 3:1225 (1961).

229. F. P. Kesamanly, D. N. Nasledov, and Yu. V. Rud', in: Physics (Proc. Twenty-Third Sci. Conf. at Leningrad Structural Engineering Institute) [in Russian], Leningrad (1965), p. 51.

230. N. A. Goryunova, F. P. Kesamanly, D. N. Nasledov, V. V. Negreskul, Yu. V. Rud', and S. V. Slobodchikov, Fiz. Tverd. Tela, 7:1312 (1965).

231. F. P. Kesamanly, Yu. V. Rud', and S. V. Slobodchikov, Dokl. Akad. Nauk SSSR, 161:1065 (1965).

232. K. Masumoto and S. Isomura, J. Phys. Chem. Solids, 26:163 (1965).

233. P. Leroux-Hugon and J. J. Veyssie, Phys. Status Solidi, 8:K60 (1965).

234. N. A. Goryunova, V. I. Sokolova, and Chiang Ping-Hsi, Zh. Prikl. Khim., 38:771 (1965).

235. G. F. Nikol'skaya, L. I. Berger, I. V. Evfimovskii, G. I. Kagirova, I. K. Shchukina, and I. S. Kovaleva, Izv. Akad. Nauk SSSR, Neorgan. Mat., 2:1876 (1966).

236. H. Fleischmann, O. G. Folberth, and H. Pfister, Z. Naturforsch., 14a:1203 (1959).

237. S. Mamaev, Izv. Akad. Nauk Turkm.SSR, Ser. Fiz. Tekhn. Nauk, 6:7 (1960).

238. H. Ehrenreich, Phys. Rev., 120:1951 (1960).

239. P. Leroux-Hugon, Compt. Rend., 255:662 (1962).

240. A. A. Vaipolin, N. A. Goryunova, É. O. Osmanov, and Yu. V. Rud', Dokl. Akad. Nauk SSSR, 160:633 (1965).

241. N. A. Goryunova and B. T. Kolomiets, Zh. Tekh. Fiz., 28:1922 (1958).

242. J. C. Woolley and B. A. Smith, Proc. Phys. Soc. (London), 72:867 (1958).

243. H. Hahn and W. Klingler, Z. Anorg. Allgem. Chem., 263:177 (1950).

244. N. A. Goryunova, V. A. Kotovich, and V. A. Frank-Kamenetskii, Dokl. Akad. Nauk SSSR, 103:659 (1965).

245. N. A. Goryunova, V. A. Kotovich, and V. A. Frank-Kamenetskii, Zh. Tekh. Fiz., 25:2419 (1955).

246.   G. Busch, E. Mooser, and W. B. Pearson, Helv. Phys. Acta, 29:192 (1956).

247.   G. A. Busch, Nuovo Cimento Suppl., 7:696 (1958).

248.   H. Hahn and G. Frank, Z. Anorg. Allgem. Chem., 269:227 (1952).

249.   N. A. Goryunova and V. D. Prochukhan, Fiz. Tverd. Tela, 2:176 (1960).

250.   J. Rupprecht and R. G. Maier, Phys. Status Solidi, 8:3 (1965).

251.   N. A. Goryunova, A. V. Voitsekhovskii, and V. D. Prochukhan, Vestnik Leningrad Gos. Univ., 10:156 (1961).

252.   A. V. Voitsekhovskii and N. A. Goryunova, in: Physics (Proc. Twentieth Sci. Conf. at Leningrad Structural Engineering Institute) [in Russian], Leningrad (1962), p. 12.

253.   S. Mamaev, D. N. Nasledov, and V. V. Galavanov, Abstracts of Papers presented at All-Union Conf. on Semiconducting Compounds [in Russian], Izd. AN SSSR, Moscow (1961), p. 41.

254.   S. Mamaev, D. N. Nasledov, and V. V. Galavanov, Fiz. Tverd. Tela, 3:3405 (1961).

255.   G. Giesecke and H. Pfister, Acta Cryst., 14:1289 (1961).

256.   H. Borchers and R. G. Maier, Metall, 17:1006 (1963).

257.   N. A. Goryunova, F. P. Kesamanly, É. O. Osmanov, and Yu. V. Rud', Izv. Akad. Nauk SSSR, Neorgan. Mat., 1:885 (1965).

258.   N. A. Goryunova, F. P. Kesamanly, É. O. Osmanov, and Yu. V. Rud', in: Physics (Proc. Twenty-Third Sci. Conf. at Leningrad Structural Engineering Institute) [in Russian], Leningrad (1965), p. 49.

259.   B. T. Kolomiets and T. F. Nazarova, Fiz. Tverd. Tela, 2:174 (1960).

260.   H. Hahn, G. Frank, W. Klingler, A. D. Störger, and G. Störger, Z. Anorg. Allgem. Chem., 279:241 (1955).

261.   F. Lappe, R. Nitsche, A. Niggli, and J. G. White, Z. Krist., 117:146 (1962).

262.   E. J. W. Verwey, in: Semiconducting Materials [Russian translation], IL, Moscow (1954), p. 201.

263.   Ferrites [in Russian], Izd. AN BSSR, Minsk (1960).

264.   L. I. Berger, in: Chemical Reagents and Materials, Trudy IREA, No. 30, p. 386 (1967).

265.   D. R. Mason and D. F. O'Kane, Abstracts of Papers presented at Fifth Intern. Conf. on Physics of Semiconductors, Prague (1960), p. 134.

266.   H. Koelmans and H. G. Grimmeiss, Physica, 25:1267 (1959).

267.   S. Shionoya and A. Ebina, J. Phys. Soc. Japan, 19:1150 (1964).

268.   R. H. Bube and W. H. McCarroll, J. Phys. Chem. Solids, 10:333 (1959).

269.   R. Nitsche and W. J. Merz, Helv. Phys. Acta, 35:274 (1962).

270.   J. A. Beun, R. Nitsche, and M. Lichtensteiger, Physica, 26:647 (1960).

271.   M. Springford, Proc. Phys. Soc. (London), 82:1029 (1963).

272.   G. Busch, P. Junod, E. Mooser, and H. Schade, Proc. Third Conf. on Physics of Semiconductors, Garmisch-Partenkirchen, 1956, publ. as Halbleiter und Phosphore, Vieweg, Braunschweig (1958), p. 470.

273.   H. Hahn, G. Frank, and W. Klingler, Z. Anorg. Allgem. Chem., 279:271 (1955).

274.   C. E. Olsen and P. M. Harris, Phys. Rev., 86:651 (1952).

275.   J. A. A. Ketelaar, Z. Krist., 87A:435 (1934).

276.   L. K. Frevel and P. P. North, J. Appl. Phys., 21:1038 (1950).

277.   D. R. Mason and J. S. Cook, J. Appl. Phys., 32:475 (1961).

278.   J. A. A. Ketelaar, Z. Physik. Chem., 26B:327 (1934).

279.   J. A. A. Ketelaar, Trans. Faraday Soc., 34:874 (1938).

280.   J. A. A. Ketelaar, Z. Physik. Chem., 30B:53 (1935).

281.   L. Suchow and P. H. Keck, J. Am. Chem. Soc., 75:518 (1953).

282.   J. C. Woolley and E. W. Williams, J. Electrochem. Soc., 111:210 (1964).

283.   O. G. Folberth and H. Welker, J. Phys. Chem. Solids, 8:14 (1959).

284.   N. A. Goryūnova, Problemy Kinetiki i Kataliza, 10:96 (1960).

285.   W. Köster and B. Thoma, Z. Metallkunde, 46:293 (1955).

286.   B. V. Baranov and N. A. Goryunova, Dokl. Akad. Nauk SSSR, 129:839 (1959).

287.   E. L. Amma and G. A. Jeffrey, J. Chem. Phys., 34:252 (1961).

288.   L. M. Foster, G. Long, and M. S. Hunter, J. Am. Ceram. Soc., 39:1 (1956).

289.   G. A. Jeffrey and H. Linton, Bull. Am. Phys. Soc., 3:231 (1958).

290.   J. C. Woolley and B. Ray, J. Phys. Chem. Solids, 16:102 (1960).

291.   N. A. Goryunova, V. S. Grigor'eva, B. M. Konovalenko, and S. M. Ryvkin, Zh. Tekh. Fiz., 25:1675 (1955).

292.   A. A. Vaipolin, N. A. Goryunova, É. O. Osmanov, Yu. V. Rud', and D. N. Tret'yakov, Dokl. Akad. Nauk SSSR, 154:1116 (1964).

293.   S. S. Shalyt, Semiconductors in Science and Technology [in Russian], Vol. 1, Izd. AN SSSR (1957), p. 7.

294.   B. F. Ormont, Zh. Neorg. Khim., 5:255 (1960).

295.   C. H. L. Goodman, Proc. Phys. Soc. (London), B67:258 (1954).

296.   N. N. Sirota, in: Physics and Physicochemical Analysis [in Russian], Gosstroizdat, Moscow (1957), p. 117.

297.   P. Manca, J. Phys. Chem. Solids, 20:268 (1961).

298.   W. Gordy, Phys. Rev., 69:604 (1946).

299.   H. Pritchard and H. Skinner, Chem. Rev., 55:745 (1955).

300.   Ya. K. Syrkin, Uspekhi Khimii, 31:397 (1962).

301.   Z. I. Ornatskaya, in: Problems in Metallurgy and Physics of Semiconductors [in Russian], Izd. AN SSSR, Moscow (1961), p. 145.

302.   N. A. Goryunova and S. I. Radautsan, in: Soviet Research on New Semiconducting Materials (ed. by D. N. Nasledov and N. A. Goryunova), Consultants Bureau, New York, (1965), p. 1.

303.   T. A. Kontorova, Zh. Tekh. Fiz., 26:2021 (1956).

304.   D. K. C. MacDonald and S. K. Roy, Phys. Rev., 97:673 (1955).

305.   L. I. Berger, in: Problems of Metallurgy and Metallography [in Russian], Metallurgizdat, Moscow (1962), p. 157.

306.   V. S. Neshpor, Fiz. Metallov i Metalloved., 7:559 (1959).

307.   M. G. Lozinskii (Losinsky) and S. G. Fedotov (Fedotow), Neue Hüte, 3:489 (1958).

308.   B. F. Ormont, Zh. Neorg. Khim., 4:2174 (1959).

309.   B. F. Ormont, Dokl. Akad. Nauk SSSR, 106:687 (1956).

310.   B. F. Ormont, Zh. Fiz. Khim., 31:509 (1957).

311.   T. A. Kontorova, in: Some Problems in the Strength of Solids [in Russian], Izd. AN SSSR, Moscow (1959), p. 99.

312.   V. V. Panchenko, Uchenye Zapiski LGPI im. Gertsena, 148:109 (1958).

313.   S. Balce, Philippine J. Sci., 60:251 (1936).

314.   E. Mooser and W. B. Pearson, Acta Cryst., 12:1015 (1959).

315.   B. T. Kolomiets and A. A. Mal'kova, Fiz. Tverd. Tela, 1:32 (1959).

316.   D. F. Edwards and D. F. O'Kane, Bull. Am. Phys. Soc., 5:78 (1960).

317.   L. I. Berger, V. I. Sokolova, and N. M. Grinberg, in: Chemical Reagents and Materials, Trudy IREA, No. 30, p. 408 (1967).

318.   A. V. Voitsekhovskii, Ukr. Fiz. Zh., 9:796 (1964).

319.   B. V. Baranov and N. A. Goryunova, Fiz. Tverd. Tela, 2:284 (1960).

320.   B. R. Pamplin, J. Phys. Chem. Solids, 25:675 (1964).

321.   B. F. Ormont, Dokl. Akad. Nauk SSSR, 129:129 (1959).

322.   C. H. L. Goodman, J. Electronics, 1:115 (1955).

323.   V. A. Presnov, Fiz. Tverd. Tela, 4:548 (1962).

324.   L. I. Berger, in: Chemical Reagents and Materials, Trudy IREA, No. 26, p. 293 (1964).

325.   D. P. Belotskii, in: Problems in Metallurgy and Physics of Semiconductors [in Russian], Izd. AN SSSR, Moscow (1961), p. 18.

326.　N. A. Goryunova, Izv. Akad. Nauk SSSR, Neorgan. Mat., 2:785 (1966).
327.　J. C. Anderson, S. K. Dey, and V. Halpern, J. Phys. Chem. Solids, 26:1555 (1965).
328.　L. I. Berger and N. A. Bul'enkov, Izv. Akad. Nauk SSSR, Ser. Fiz., 28:1100 (1964).